数字工厂
高级计算性生形与建造研究
DIGITAL FACTORY
ADVANCED COMPUTATIONAL RESEARCH

学生建筑设计作品
DADA 2015
STUDENTS

徐卫国 / 尼尔·林奇（英）编
Xu Weiguo / Neil Leach [eds.]

中国建筑工业出版社

目录 / CONTENTS

前言	6
PREFACE	6
对话	8
DIALOGUE	8

America

美国加州艺术学院	22
CCA	22
美国哥伦比亚大学	28
Columbia GSAPP	28
美国哈佛大学设计研究生院	34
Harvard GSD	34
美国麻省理工学院	40
MIT	40
美国普瑞特艺术学院	46
Pratt	46
美国普林斯顿大学建筑学院	52
Princeton	52
美国伦斯勒理工大学	58
RPI	58
美国南加州建筑学院	64
SCI-Arc	64
美国加州大学洛杉矶分校建筑系	70
UCLA	70
美国密歇根大学	76
UMichigan	76
美国宾夕法尼亚大学建筑系	82
UPenn	82
美国南加州大学建筑学院	88
USC	88
美国耶鲁大学建筑学院	94
Yale	94

Europe

英国建筑联盟建筑学院	100
AA	100
奥地利维也纳工艺美术学院	106
Angewandte	106
英国伦敦大学巴特利建筑学院	112
Bartlett	112
丹麦皇家美术研究院信息技术与建筑中心	118
CITA	118
荷兰代尔夫特工业大学	124
TU Delft	124
德国德绍建筑学院	130
DIA	130
欧洲研究生院	136
EGS	136
西班牙加泰罗尼亚高级建筑研究学院	142
IAAC	142
法国巴黎玛莱柯建筑学院	148
Paris Malaquais	148
德国斯图加特大学	154
ICD Stuttgart	154
瑞士苏黎世联邦理工大学建筑学院	160
ETH Zurich	160

Australia

澳大利亚皇家墨尔本理工大学	166
RMIT	166

China

中国香港中文大学建筑学院	172
CUHK	172
中国香港大学建筑学院	178
HKU	178
中国南京艺术学院	184
Nanjing UA	184
中国华南理工大学建筑学院	190
SCUT	190
中国东南大学建筑学院	196
Southeast	196
中国天津大学建筑学院	200
Tianjin	200
中国同济大学建筑与城市规划学院	204
Tongji	204
中国清华大学建筑学院	210
Tsinghua	210

索引	216
INDEX	216
作者简介	222
BIOGRAPHIES	222

前言 / PREFACE

本作品集收录了在上海同济大学建筑与城市规划学院举办的"数字工厂：高级计算性生形与建造研究"学生作品展中的作品，该展览为DADA2015系列活动之一，其余的活动包括一系列的工作营与国际会议等。

自2004年起，徐卫国与尼尔·林奇合作策划了一系列的数字建筑展览。第一次是2004年在北京UHN国际村为北京建筑双年展举办的"快进>>"展。之后4次分别是2006年在北京世纪坛举办的"涌现"展，2008年和2010年在北京798时态空间举办的"数字建构"展和"数字现实"展，以及2013年在北京751举办的"设计智能：高级计算性建筑生形研究"展。本次展览是往届合作策展的延续。

本作品集收录了一些世界顶尖建筑院校最高级的计算性设计作品。这些学校包括英国建筑联盟建筑学院、哈佛大学设计研究生院、麻省理工学院、斯图加特大学、伦敦大学、清华大学和同济大学。至此，本次活动呈现出各个建筑院校持续增长的对于计算性建筑研究的兴趣爱好，并且提供了非常有用的该领域近期发展的剪影，与先前的活动共同展现出该领域的发展全貌。

This is a catalogue of works on display in the 'Digital Factory: Advanced Computational Research' exhibition of student work at the College of Architecture and Urban Planning, Tongji University, Shanghai, as part of the DADA 2015 series of events. Other events in DADA 2015 include a series of workshops and a conference.

Since 2004 the curators, Xu Weiguo and Neil Leach, have collaborated on a series of digital design exhibitions. The first exhibition, 'Fast Forward>>', took place in UHN, Beijing in 2004 as part of the Architecture Biennial Beijing. This was followed by four further exhibitions, 'Emerging Talents, Emerging Technologies', in the Millennium Museum, Beijing in 2006, '(Im)material Processes: New Digital Techniques for Architecture' in 798 Space in 2008, 'Machinic Processes' in 798 Space, Beijing in 2010, and 'Design Intelligence: Advanced Computational Research' in 751 D-Park, Beijing in 2013. This exhibition is a continuation of that collaboration.

This catalogue offers a showcase of the most advanced computational design work from some of the leading schools of architecture in the world. These schools include the Architectural Association, Harvard University GSD, MIT, University of Stuttgart, The Bartlett, Tsinghua University and Tongji University. As such, it serves to track the continued growth of interest in computational research in schools of architecture, and provides a useful snapshot of recent developments in the field that adds to the collective overview provided by previous exhibitions.

2015年的展览有一个非常重要的特征,那就是机器臂建造受到极大的关注。实际上,这种情况不仅仅发生在早已有机器臂的那些学校;在中国,像清华大学和同济大学这样的顶尖院校,现在也配置了机器臂;这说明中国大学的建筑专业已处于世界建筑学发展的前沿。

最后,策展人对所有为作品集做出贡献的人致以最诚挚的谢意。特别感谢刘畅、张鹏宇、刘春、刘洁、秦承祚、金旖、张自牧、刘剑颖、张弛和时思芫等为作品集的设计、翻译和材料整理所做的工作;同时,对提交作品的院校致以谢意。

尼尔·林奇
徐卫国

One significant feature of this year's exhibition has been the increasing popularity of robotic arms used in fabrication. Indeed it is not only the more established Western schools that have acquired robots. Leading schools in China, such as Tsinghua University and Tongji University, have also now acquired them, illustrating the remarkable speed at which Chinese schools of architecture have been catching up with the rest of the world.

The curators are grateful to all those who have contributed to the preparation of this catalogue. In particular, they would like to thank Liu Chang, Zhang Pengyu, Liu Chun, Liu Jie, Qin Chengzuo, Jin Yi, Zhang Zimu, Liu Jianying, Zhang Chi and Shi Siyuan for their invaluable contribution in helping to design, translate and compile material for this exhibition, and to the schools themselves for submitting that material.

Neil Leach
Xu Weiguo

对话 / DIALOGUE

徐卫国—尼尔·林奇
XWG-Neil Leach

徐卫国（以下简称徐）：去年深圳 DADA 核心会员会议决定2015年系列活动由同济承办，袁烽负责，并确定活动主题为"数字工厂"。我想这一主题包含了多层含义：其一，DADA 试图推动数字技术从数字设计向数控生产发展；其二，建筑业应该以数控加工作为产业升级的目标；其三，建筑的数控建造应该能响应新兴工业4.0的崛起，并携手探索智能建筑工业的道路；其四，DADA 有责任召唤建筑师站在新一轮工业革命的前列，引导数字建筑产业的诞生及健康发展。这是"数字工厂"这一命题的含义。您对这个题目作何想法？

Neil Leach（以下简称 NL）：我认为这个题目非常适合一个位于中国的展览。首先，所有人都知道目前世界上很多商品都是中国制造的。中国已经成为了世界工厂，因此在中国举办这个展览非常合适。但是在此背后，这个题目试图传达的是对于实际生产的关注。工厂毕竟是一个主要追求生产效率最大化的实体，中国作为建筑行业中的领先大国，在数字加工的领域也产生了越来越多的对于数字技术与实际生产相结合的关注。例如，像 E-Grow 这样的公司已经出现在我们的视野中有一阵了。E-Grow 是西方领先的建筑公司，为扎哈·哈迪德及墨菲西斯在中国的项目提供数字生产的面板系统。同样，我们也可以引用最近在中国三维打印出来的住宅作为一个例子，尽管它所使用的技术仅仅是由南加州大学克什涅维斯教授所发明的轮廓工艺的简化版本，但是在中国尝试此类项目本身仍然具有很大意义。另外，中国目前也有计划依照克什涅维斯教授的技术制造原型机。现在 3D 打印现象随着建筑师跨界到其他创意领域已经蔓延至整个中国。例如，由 OMA 创始人雷姆·库哈斯的侄子雷姆·D·库哈斯创立的 United Nude。它目前在广州并与 UN Studio、扎哈·哈迪德等前卫的设计师事务所合作生产 3D 打印的鞋子。与此同时设计师马子聪也在上海创建了自己的事务所，专注于设计三维打印的首饰及可穿戴物。

目前，数字加工在中国学术圈内同样扮演着越来越重要的角色。机械臂也被引入到中国的若干所建筑院校之中。就像它转变了一些西方学校的教育模式一样，它也同样改变了中国的建筑教育。同

XWG: Last year, Tongji University was selected to host DADA 2015 event series in the core members meeting of DADA. The theme for the DADA 2015 events was decided to be "Digital Factory". There are many layers of meanings related to the theme. First, DADA is promoting the focus of digital technology to migrate from digital design to digital production. Secondly, digital production should be the direction of the future upgrade of building industry. Also, under the inevitable rising of industry 4.0, the CNC building construction will take on the road to discover the future of intelligent building industry. Lastly, DADA should be responsible for calling architects to make efforts to lead the development of digital building industry. What do you think about the title?

NL: I think that it is a very good title for an exhibition opening in China. To begin with, everyone knows that China produces many of the goods in the world right now. As such, China has become the world's factory, and it therefore seems very appropriate to hold this exhibition in China. But – beyond this – the title seems to convey a certain focus on practical production. Factories, after all, are primarily practical spaces dedicated to efficient production. This interest in practical productions itself reflected in the digital fabrication processes that are playing an increasingly significant role in what has also become one of the leading countries in the world in terms of building construction.For example, we have known about companies like E-Grow for some time now. E-Grow has supplied digitally fabricated paneling systems for leading Western architects operating in China, such as ZahaHadid Architects and Morphosis. We might also cite the example of the houses that were 3D printed recently here in China.While the technique used was essentially a simplified version of the Contour Crafting technique invented by BehrokhKhoshnevis, a professor of engineering at the University of Southern California, it is nonetheless significant that there here in China is the will to attempt such a project. Also there are now plans to prototype ProfessorKhoshnevis's technique in China.Indeed the phenomenon of 3D printing has now begun to spread throughout China as architects become involved increasingly in other creative industries.For example, United Nude – the company set up by Rem D Koolhaas, nephew of the more famous founder of OMA – is now operating out of Guangzhou, producing 3D printed shoes in collaboration with several leading architects, including UN Studio and ZahaHadid Architects, while Steven Ma has opened up an office in Shanghai, focusing on 3D printed jewelry and wearables.

Meanwhile digital fabrication is also playing an increasingly significant role in Chinese academic

时，DADA 的队伍也越来越壮大。2015DADA 数字工厂研习班将一批极具天赋的教师聚集在了一起。事实上，我甚至不知道任何一所西方的学校可以如此自豪地引入一批世界级的教授来研习班授课。令人瞠目结舌的是这里同时有阿希姆·门格斯、罗兰·斯努克斯、马德朴、约翰·布朗曼以及雷姆·D·库哈斯这样的西方的领军人物及中国的代表袁烽、于雷、宋刚、马子聪等并肩教学。

另外，数字未来系列图书的最新一本《人机未来》出版了，并在上一本《制造未来》的基础上带来更进一步的贡献。所有的这些进步都表明了不仅仅在建筑建造中，更是在学术及研究领域，中国已经成为数字生产制造界的中心。中国已经成为建筑生产领域新的数字工厂了。

徐：那么，总体而言，在数字设计的发展方面，如果您把中国和西方进行对比，将得出什么结论？

NL：我记得当 2004 年，我们共同策划第一个展览的时候，我们为它取名为"快进>>"，这个名称来源于在录音播放机上用于加快播放的按钮。但是那时候中国的建筑文化同计算机的交会很少，甚至说那时候几乎在中国建筑界，大家从未听说过编程，所有的事情都很原始；但是到了我们上一个 2013 年的"设计智能：高级计算性建筑生形研究"展览的时候，中国已经发展出了非常成熟的数字设计文化。实际上，我甚至能想起当马克·贝瑞（他曾受邀参加我们 2004 年第一届的展览）坐在我们 2013 年的 DADA 会议现场的第一排时，他看上去被当时的所见所闻震惊得合不拢嘴。他最后转过身来对我说："我要回到澳大利亚告诉我的同事们，我们已经没有任何东西可以教给中国建筑师了。"我不太确定他的话是否有一定的夸张，但是这确实导致我反思我们最早的题目"快进>>"以及不同的快进速度。当我们提出这个题目的时候，我脑海中是 4 倍甚至 8 倍快进的速度，但是事实上我们观察到的却是 16 倍甚至 32 倍的快进速度。在中国，数字设计的追赶及发展速度令人震惊。

在 2004 年之后，另外一个中国发展的分水岭是 2008 年。这一年的标志是为了北京奥林匹克运动会，鸟巢及水立方两个代表性的

circles. Robotic arms have now been introduced into several schools of architecture here, thereby transforming Chinese architectural education in a way that it has already transformed several schools in the West. Meanwhile, the DADA initiative has gone from strength to strength. The 2015 DADA Digital Factory workshops are remarkable for the range of talented instructors that they bring together. In fact I don't know of any Western school that could boast of the same array of world-class professors teaching workshops. To have leading Western figures, such as Achim Menges, Roland Snooks, Matias del Campo, Johannes Braumann, and Rem D Koolhaas, running workshops alongside leading Chinese figures, such as Philip Yuan, Yu Lei, Gang Song, Steven Ma and so on, is an astonishing development.

Moreover the publication of Robotic Futures, - the latest in the Digital FUTURE series of books that builds upon our earlier publication, Fabricating the Future – makes a further contribution. All these developments only demonstrate– not only in terms of building construction, but also in terms of both pedagogy and academic research – that China is now playing a central role in the development of these digital fabrication techniques. China has become the new digital factory for architectural production.

XWG: These changes in education will help the rapid forwards in 'Digital Factory' in China. Generally, how do you think in the developments of digital design, if comparing china to the West?

NL: I can recall that back in 2004 when we curated our first exhibition we called it 'Fast Forward>>', as a reference to the control button on most audio-visual systems that allows one to increase the speed of that system. But in those days Chinese architectural culture had hardly engaged with computation. Indeed it seemed that no one in architectural circles in China had even heard of scripting. Everything was very primitive. But by the time of our last exhibition in 2013, Design Intelligence: Advanced Computational Research, China had already begun to develop quite a sophisticated digital design culture. Indeed I can recall seeing Mark Burry – who had been invited to that first exhibition back in 2004 – sitting in the front row of our 2013 conference, and it looked as though his jaw dropped in incredulity at what he was seeing. He subsequently turned to me and said, 'I have to go back to Australia and tell my colleagues there that there is nothing that we can teach these Chinese architects any more.' I'm not sure if this is strictly true, but it did cause me to reflect on our original title, 'Fast Forward>>', because in fact there are many different rates of 'fast forward'. I had been thinking in terms of 4 times, or even 8 times as fast.

计算机设计的建筑的诞生,同时期还有OMA的CCTV大楼。在鸟巢的设计当中应用了Digital Project软件,即第一个建筑信息建模技术的案例;其中,Digital Project被用来控制每条放样曲线的定位,并通过将所有的建造信息融合进一个数字模型里的方式来控制造价以及建造周期。另一方面,很明显的变化是建筑设计事务所开始大量投入自己的计算机研究团队。例如诺曼·福斯特事务所的特殊建模团队(SMG)、奥雅纳公司内部的高级几何团队(AGU),以及扎哈·哈迪德事务所的CODE小组等。实际上,奥雅纳作为工程咨询公司,参与了上面提到的两个项目;AGU小组的资源也得到了充分的利用。同样的,盖里科技也作为顾问参与到了鸟巢的设计建造之中。现在,计算机的应用已经是重大工程必不可少的一部分了。我之前提到的关于工厂的参考,实际上它目前已经成为实际生产中必不可少的一环。同时,计算机工作营与从前仅仅利用计算机技术将传统的设计过程数字化不同,现在的工作营尝试利用电脑技术生成设计。目前中国有非常多类似的工作坊,他们吸引来了大量的中国学生,袁烽和我甚至决定出版一本叫做《数字工作营在中国》的书。2008年注定了计算机技术将在中国扎根。

徐:刚才你用2004年我们策展的"快进>>"展览题目来形容中国数字设计的发展,这非常形象、非常确切。事实上,近十年来,中国已经形成了数字建筑设计师群体,我们称其为"数字新锐",他们都有实际建成项目,这些建成项目充分依靠和展现了数字技术的潜能。比如北京市建筑设计研究院邵韦平设计的凤凰卫视北京媒体中心,在设计完成度、设计的细节、设计的过程逻辑、材料的使用,以及结构的合理性、构件加工及施工的精确度等方面都可以说达到了世界最好,大家更喜欢凤凰卫视的亲切及细腻,我甚至把它与哈尼·拉什德的阿布扎比YAS酒店进行了比较(去年有机会入住YAS酒店),凤凰卫视要比YAS酒店好很多。此外,张晓奕设计的阿里巴巴杭州总部大厦、林秋达设计的厦门第三国际航站楼、宋刚设计的佛山艺术村、王振飞设计的青岛园博会服务设施建筑等,都是数字技术的成果,中国宏大的建设规模以及开放的社会心态使这些年轻建筑师有机会在实践中展示才华,他

In fact I think that what we have witnessed has been in the order of 16 or 32 times as fast. It is astonishing how quickly digital design has caught on in China.

After 2004 the next watershed moment – at least as regards China – was 2008, a year that marked the completion of not only two highly computationally designed buildings for the Beijing Olympics, the 'Birds' Nest Stadium' and the 'Water Cube', but also the construction of the OMA's CCTV building. On the one hand, Digital Project – one of the first examples of Building Informational Modelling (BIM) – had been used in the Bird's Nest Stadium to control not only the precise definition of the curves employed, but also to control the cost and time of construction by integrating all construction information on a single digital model, that could be updated relatively easily. On the other hand, it was clear that architectural offices were beginning to invest heavily in in-house computational research units, such as the Specialist Modeling Group (SMG) at Foster and Partners, or the Advanced Geometry Unit (AGU) at Arup, or CODE at ZahaHadid Architects, and so on. In fact Arup was the consultant engineering firm for both projects, and the resources of the AGU were used extensively. Likewise Gehry Technologies were involved as consultants for the Bird's Nest stadium. By this time we had reached the stage that computation had become seemingly indispensable in any major construction project. In terms of my earlier reference to factories, it had become a practical necessity in terms of production. Meanwhile computational workshops – that sought to harness the capacity of the computer to generate designs, rather than simply as a 'computerization' of the culture that existed before the introduction of computers – began to spread like a virus all over China. Indeed there were so many computation workshops attended by so many Chinese students, that Feng Yuan and I decided to publish a book about the phenomenon, Digital Workshops in China. By 2008 it was clear that computation was here to stay.

XWG: You have just used the word "Fast Forward>> " in the exhibition of 2004 we curate together to descript the development of digital design in China. This is very accurate. In fact, in last decade, a new group of digital architects are emerging, called "Digital Advanced". They all have practical projects realized relying on the digital technology. For example, the media center for Phoenix TV in Beijing designed by Shao Weiping of Beijing Institute of Architecture Design, many aspects including the level of completion, design details, logic design process, the selection of materials and structure system, component fabrication and construction accuracy have reached the top level in the world. People like the exquisy and friendliness of Phoenix Center. I have even compared it to the YAS Hotel in Abu Dhabi by Hani Rashid as I had the stay there last year, and

们将引领数字建筑设计的未来。

NL：但是，对于数字设计及生产建造本身而言，过去几年曾有过一个概念上的巨大转变。在2004~2008年这段实践中，对于数字工具的使用主要是在外形及表现上，计算机大多数时候被用于作为一个产生新的美学形式的工具，在那之后我注意到了关注点由表现形式到过程及性能的转变。我先要用"形体"的词根（form）来讨论。我们一方面可以用"为了形体而找形"又或者我们可以从更高、更成熟的角度将形体（form）延展到以它为词根的单词上，例如"性能"、"活跃度"及"信息"。这样，问题就变成了"表现"如何能够"影响"到"形体"，这在中国虽然并不是一个随处可见的设计浪潮，但是数字设计界的一些带头人已经开始向这些方向转变；同时，在西方由性能引导设计方向的带头人阿奇·蒙哥斯也在中国越来越有影响力。

徐：是的，这一点在建筑设计教学上表现得尤为突出，比如清华的数字设计工作坊已有十年的历史。开始的时候侧重于用软件和算法生成形态，但现在已完全基于建筑的性能条件，如环境因素、人的动态行为，以及相互间的互动等等，并在寻找基本联系的前提下运用参数模型进行设计生成，其结果可以直接满足设计需求并与环境之间建立友好的联系。对这些学生进行训练，使他们建立起更科学的建筑设计观；而当他们毕业工作后，他们新的设计思想又影响了设计实践；有一些学生会去西方留学，他们也将影响到西方的建筑教学。

换一个话题，我们来谈谈"风格"的问题，之前您似乎谈到过，用"风格"来评判数字设计是不合适的，能否再谈谈对这一问题的看法？

NL：不，完全不合适。我认为将计算机设计定义为一个建筑风格是错误的，我不同意舒马赫所言的有一个名为参数化主义的全球性建筑风格，并且其特点是曲面建筑。我在2010ABB会议纪要上曾经写过一篇文章，深入探讨了"参数化主义"的定义及技术上的误用，在我们当今的大环境之下讨论这种全球通用的概念已经没有太多的意义了。在现代主义倒塌之后此类的概念与利奥塔提出后现代主义中"辉煌叙事"的倒塌相背驰。同样的，对于风格

found that the Phoenix center is much better. Also the Alibaba headquarter in Hanzhou by Zhang Xiaoyi, Terminal 3 of Xiamen Airport by Lin Qiuda, Fushan Art Village by Song Gang, the service facilities of Qingdao Gardening Expo are all the examples of application of digital technology. The massive construction going on in China and the open minded society provide opportunities for these young architects to reveal their talent in practice. They will lead the future of digital architecture design.

NL: But I think that there has also been a significant shift in terms of the nature of digital design production. In the 'old days' – if we can refer to a period as recent as 2004-2008 in that way – the interest in digital tools operated primarily at the level of representation. The computer was largely a tool for producing a new aesthetic. What I have noticed since then has been a shift away from mere representation towards process and performance. I like to think of this in terms of the word 'form'. We can either think of 'form' in terms of 'form for form's sake'. Or we can think of it at a more sophisticated level in terms of a series of other terms that also include the word 'form', as in 'performance', 'performativity' and 'information'. The question now is how issues such as 'performance' can 'inform' that 'form'. This is by no means a universal phenomenon in China, but some of the leading figures in digital design, are beginning to think more in these terms, and the work of AchimMenges, who has been a pioneer of performance driven design in the West, has come to be recognized as highly significant here in China.

XWG: Yes, This is very explicit especially in architecture education. For example, the digital design studio in Tsinghua University have been ongoing for 10 years. At the earlier stages, it was more focused on software application and algorithm form generation. But now it has been based on the architecture performance, such as environmental factor, human behavior and interaction etc. And the parametric models are based on those basic connections among different factors. The result of the generating will directly bridge the design requirements and the environments. This makes students establish a more scientific design method. After graduation, their design method will in turn influence design practice. Some of the students who will be studying abroad have brought new knowledge to western education.

For the issue of "Style", do you think of digital design as being a question of 'style'?

NL: No. Not at all. I think that it is a mistake to think of computation in terms of a new 'style' for architecture. I certainly disagree with Patrik Schumacher, who claims that there is a new global style of architecture, called 'Parametricism', that is characterized by its curvilinear forms. I have

等问题的过度关注又貌似是一个后现代主义的趋势，因此，舒马赫被困在现代主义和后现代主义之间。同样，舒马赫将计算机出现之前所生成的形体同计算机出现之后所生成的形体混为一谈，由此将曲线形体的出现完全归功于计算机的使用也是毫无意义的。这有些类似于人们曾经攻击计算机设计的肤浅、表面性及缺少内涵，而且我们已经在如《向拉斯韦加斯学习》这类出版物中看到了现有的整个由肤浅和表面性构成的文化。

对我个人而言，计算机仅仅是个工具，同时也是建筑想象力的延展。就它本身而言并没有固定的倾向，因此它并不能导致建筑师以某些特定的方式工作。但正如J·吉普森所言，它具有一定的易上手性，即通过它的使用使设计师得以更加便捷地做出一些操作。例如，以前的平行尺几乎不可能绘制出一条曲线的控制线。当我在剑桥读书的时候，我们使用的是一种"法式曲尺"，即一种可以弯成曲线的软尺，但是我们没办法确切地理解这条曲线背后的逻辑。约恩·伍重在设计建造悉尼歌剧院的时候，结构顾问为奥雅纳，为了能够有效理解屋顶的自由曲线，不得不将其简化为一系列的球面。当然，在计算机出现之前我们也能创造曲线的造型，区别仅仅是现在通过计算机的应用，我们能够更简单地控制曲面，曲面形体的数字建造也是类似的情况。回想起2006年，我们曾经试图依照电脑画出的图纸，靠手工工具加工出AAB 2006的展亭，而在理想情况下，展亭应当用计算机数控机床（CNC）来加工，但是那时候在中国找不到这样的机器，于是我们使用了大量的建筑工人来手工切割出所需形体构件。最终，就像上面所提到的，控制的问题延伸到了整体的施工当中，其中BIM技术颠覆了建造过程中预算及工期控制的逻辑。

计算机的使用可以使类似的工作更加便捷。由此，我们可以观察到数字设计受到了一些共同影响，但是本质上，我不认为计算机的使用导致了一种新的风格的产生。与其说是风格，不如说计算机的使用使得人们可以更加轻易地做出一些特定的操作。例如，当今在线订机票远远比去中介公司更方便，结果便是计算机的普及导致了航空业的发展。更进一步来说，人们甚至可以将由越来

gone into the technical misunderstandings of the term, 'Parametricism', in my essay in the 2010 ABB catalogue. Also it makes no sense in our contemporary condition to talk about universal concepts.Following the collapse of Modernism such ideas are incompatible with what Jean-Francois Lyotard has called the collapse of 'grand narratives' in our Postmodern age. Equally to focus on the question of style seems to be a highly postmodern tendency.SoPatrik is trapped somewhere between Modernism and Postmodernism.ButPatrik also confuses the forms that were generated before the introduction of computation with those that were produced afterwards, and – as such – it makes no sense to attribute curvilinear forms purely to the introduction of the computer. It's a bit like those that attack computation as being superficial, shallow and meaningless, when we already saw the rise of a culture of superficiality with the publication of Learning from Las Vegas.

Personally I see the computer as just a 'tool', but a very sophisticated tool that becomes an extension of the architectural imagination. In itself it has no agency, and therefore cannot cause an architect to operate in any particular way.But it has certain 'affordances' – as J JGibson would put it – that makes it easier for the designer to perform certain operations. For example, in the old days of parallel motion drawing boards it was almost impossible to control the line of a curve. When I was a student at the University of Cambridge we used to use a 'French Curve' – a kind of flexible ruler that could be bent in the form of a curve – to describe them.But we had no way of knowing precisely the logic of that curve. Indeed it is significant that for JornUtzon's Opera House in Sydney the consultant structural engineers, Arups, had to reduce the freeform curves of the roof to a series of segments of spheres in order to understand them.Of course, we had curvilinear forms before the computer.It's just that it is easier to control curvilinear forms using a computer. And the same goes for constructing curvilinear forms using digital fabrication. As you recall, in 2004 we attempted to manually fabricate a pavilion for AAB 2004 based on computational drawings. Ideally the pavilion should have been fabricated using a Computer Numerically Controlled (CNC) milling machine, but we did not have access to one in China at that time, and instead employed an army of construction workers to manually cut the forms. The result was a disaster.Finally, as mentioned above, the issue of control extends to the overall construction, in that BIM has revolutionized the logistics of construction in terms of controlling cost and time.

So the computer makes it easier to perform certain operations. As such one can maybe recognize certain common affects in some digital designs. But – fundamentally – I do not think that the

越多的飞行里程导致对环境的破坏归结到计算机的使用上。但是归根结底,选择乘坐飞机出行的人才是环境危害的缔造者,我们不能将其归罪于计算机。就像我之前提到的,电脑本身没有倾向,它仅仅是一个工具,一个人类想象力延伸的假肢。

徐:我同意你的观点"计算机设计不能被定义为一种建筑风格",但是,难道你没有注意到,由于计算机的出现,建筑设计出现了一类造型——那是如果不用计算机就不可能产生的造型?我们又怎么解释这种设计现象呢?

我的观点是,一方面,数字技术正无孔不入地渗透到建筑设计的方方面面,并导致其发生激变,这种渗透的结果提高了现有建筑设计的效率和质量。比如对于标准几何形体的生形及控制变得轻而易举,比如设计的精确度大大提高等,这是数字技术工具性的体现。数字渗透的另一结果是解决了过去那个时代遗留下来的许多症结,比如以BIM模型及协同设计为基础的设计组织方式使得设计团队里不同建筑师的局部工作得到整体统合,使得不同专业在设计阶段的矛盾可以被及时发现,并消灭在投入施工之前等,这是数字技术特有的功效性。同时,数字渗透实现了建筑师过去的许多建筑理想,比如生态建筑在数字技术的支持下得到了不同程度的实现,性能模拟、环境响应、室内舒适度数控等,均是建筑生态节能的表现;场所及场所精神理论追求建筑、人、环境三者之间的互动及其特殊的场所感,互动建筑技术可以真正实现三者的互动,并使人具有个性化的归属感;建构理论推崇建筑形式忠实地表现建筑的结构及构造逻辑,而算法生形及数控加工可以最大程度地实现形式与结构系统及材料构造逻辑之间的对应,这些都是数字技术运用于现有建筑设计的结果。

另一方面,数字技术还催生了一个新的建筑设计趋势,其结果展现了前所未有的数字新建筑图景,它确实以新的造型作为标签,标示出了建筑设计的新动向。当然,其内涵实际上大大超过了形式本身,数字技术改变了建筑师的设计思考方式,比如寻找"关系"及"规则"是设计的出发点,算法是设计的核心内容,这与传统的设计思维有着天壤之别。数字技术还改变了设计过程,比如它

computer causes a certain style to emerge. Rather it just makes it easier to perform certain operations. For example, it is much easier to book a flight on line these days, rather than visit a travel agent. As a result it might be that the use of computers leads to an increase in air travel, and some might even blame the computer for causing environmentally unsustainable practices such as frequent air travel. But – ultimately – it is the person booking those flights that could be accused of being unsustainable. We cannot blame the computer. As I mentioned above, the computer has no agency. It is simply a tool that becomes a prosthesis to the human imagination.

XWG: I agree with your opinion that "Computer aided design cannot be defined as an architecture style". But perhaps you have noticed, due to the emerging of computer, there are architectural forms appearing that were made possible only through the application of computer. How do you explain this scenario?

From my viewpoint, in one hand, digital technology is infiltrating into all aspects of architecture design and causing dramatic changes in the field. The result of Digital Infiltration drastically improves the efficiency and quality of current architecture design. For example, the form-finding and form-controlling of standard geometry would be more accessible, the accuracy of design would be greatly improved. Another result of Digital Infiltration is to resolve the leftover issues of the past era. For example, the BIM system and cooperative design would integrate the works of different architect in the project, providing the ability to eliminate the conflict among different specialties before the project goes into construction. In addition, Digital Infiltration could help realizing the architect's fantasies of the past age. For example, eco-design with the support of digital technology can be realized at a different level including the performative simulation, environmental response, interior comfort control and so on. The theory of Place and Genius Loci seeks interaction between Building, Human Being, Environment and their positive interaction. Interactive Design could really achieve the interaction of the three elements and reinforce the sense of belonging for the user. The spirits of Tectonics is that architectural form should be the presentation of its logics of structure and material. The algorithm design and CNC fabrication could reach the spirits of Tectonics at the highest level. These are the result of applying digital technology onto the current architecture design.

On the other hand, digital technologies are creating totally different architectural scenery from current design. It indeed takes a new kind of form or style as label to reveal a new tendency. In fact, it's influence is exceeding its form and becoming greater and deeper. It transformed the

再也不是建筑师发挥灵感的形式创造过程,而变成了基于设计需求,通过构筑参数模型反复求解的形式搜寻及形式优化过程。无可置疑,上述这些变化将缔造一种数字新建筑,它所包含的内容远远不止建筑的形式风格。前者是改良性的,而后者是创新性的,但它们都得益于计算机科学技术的支持,同时推动建筑设计向着更科学的方向发展。

到目前为止,我们已经谈到了关于生产"设计生形"以及"建筑建造";但是作为一个理论家,你如何从理论的高度来描述数字设计呢?

NL:在我的印象里,我们目前还没有一个对于计算设计的全面理论。一方面,我们有一系列由"极客"们撰写的技术性的评论文章,他们非常了解计算技术,但是对理论却不甚了解。另一方面,曾有几个作者试图从旁观者的角度撰写计算设计。后者包括历史学家如安东尼·皮康和马里奥·卡波,他们从历史学的角度来观察计算设计的发展,但是却几乎同计算设计领域没有任何直接的接触,因为他们不是设计师。

比如安东尼·皮康从技术史的角度出发,而马里奥·卡波则是从文艺复兴和巴洛克的角度出发。如果让我自我批判的话,我甚至可以说我太倾向于从大陆哲学的角度来评判计算设计。事实上,如果你们读过我之前发表在 ABB 论文集中的文章的话,会发现我过于频繁地通过类似于吉尔·德勒兹的视角来评论计算设计。但是他本人完全不了解计算设计,因为他在 1995 年便过世了,那时计算机尚未在文化生活中产生巨大的影响。尽管我也亲自带过一些最为前卫的计算机设计课,并且参与了一些前沿的计算机研究项目,例如美国国家航空航天局 NASA 赞助的在月球和火星上使用机器臂建造技术的项目,但目前我也处于同样的境遇中。

我们同样可以认为理论必须来源于实践,而且实践需要一段时间的沉淀与发酵才能酝酿出关于计算机的理论。类似的,我们也需要涉及另一个观点即"理论已死",并且这个观点在一些圈子里面越来越流行了。当然,我们可以说理论的高峰已经过去了(在 20 世纪 90 年代理论非常流行),甚至就我的观点来说,"理论已死"的说法本身便是一个理论的悖论;但无论是什么原因,目前看来

way architect's thinking, for example, seeking for the 'relationship' and the 'rule' is the original point of design, algorithm becoming the core of the design. It also transformed design process, for example, the form-finding is no longer from architect's inspiration, but the result of generating and optimizing by calculation with the parametric model. Indisputably, all those changes will create a new type of architecture. No matter contents or forms of this Digital New Architecture will absolutely be different from current architecture. The former is incremental while the latter is innovative, but they both benefit from the support of digital technology and propel the architecture design to take a more scientific route.

So far, we have talked about production – the digital generation of designs and the digital production of buildings. But as a theorist, how do you theorize digital design?

NL: To my mind, we simply do not have a comprehensive theory of computation as yet. On the one hand, what we have are a series of technical commentaries, written by 'geeks' who really understand computation, but do not have a strong grasp on theory as such. On the other hand, there have been several individuals who have written about computation from an outsider's perspective. This last category would include historians, such as Antoine Picon and Mario Carpo, who read computation through a largely historical lens, but who have had almost not direct interaction with the field of computational design, because they are not designers.

For example, Antoine Picon approaches the question from the perspective of an historian of technology, while Mario Carpolooks at it through the lens of the Renaissance and the Baroque. If I were to criticize even myself, I would say that I too have a tendency to view computation largely through the lens of Continental Philosophy. Indeed, if you look at various essays that I have published on computational design for the various ABB catalogues, you will see that too often I look at computation through the lens of figures like Gilles Deleuze, who himself had absolutely no understanding of computation because he died in 1995, before computation had really taken any significant hold on cultural life. And so even though I have a considerable amount of hands-on experience of leading computationally based design studios, and of working on some fairly advanced computational research projects, such as the NASA funded projects to design a robotic fabrication technology for the Moon and Mars, I am also in the same situation.

We could also argue that theory necessarily comes after practice, and that it is going to take some time before any theory of computation can be written. Likewise we might even point towards certain claims about the 'death of theory' that have become popular in certain circles,

我们还没有一个完整的数字设计的理论体系。

对我而言，挑战在于这个理论必须来源于计算设计领域内部，并且会影响我们如何看待现实世界。在此我想要引用的一个案例是雷诺兹·克雷格所研发的用来模拟鸟群行为的集群算法模型。为了建立这个行为的模型，雷诺兹必须先为其建立一些规则，仅仅是这些规则建立之后，生物学家才得以真正地理解现实中鸟类的行为。碰巧几周前我在欧洲研究生院中听到约翰·弗雷泽介绍的理论也有类似的潜质。他是世界上数字设计领域最伟大的专家，曾经撰写过知名的著作《建筑的进化》，他曾和约旦帕斯克以及塞德里克普赖斯合作参与了"趣味宫"项目，并且现在开始拥护数字设计的理论，这令我非常惊讶。

徐：技术性的评论可以帮助建筑师辨析和提高设计技巧，但上升不到设计理论的高度，从技术史或建筑史的角度阐述计算机数字建筑设计似乎还不到时候；但我认为从德勒兹哲学出发，论述数字设计为建筑师奠定了设计的思想基础，特别是为数字建筑设计过程提供了抽象的参照系，我们应该深入系统地阐述这一理论并使它成为完善的数字建筑设计理论。

另一方面，复杂科学理论是数字建筑设计的科学基础，将复杂科学的理论及方法通过模型算法及编程机制应用于设计过程，这将大大拓展数字设计的疆域，目前已经在复杂适应性系统方面有所研究，如元胞自动机、多智能体、遗传算法、复杂网络等，但在非线性系统理论如分形理论、混沌理论、形式语法、人工神经网络等方面还有广泛的研究前景。复杂科学理论的研究将引发数字设计思维方式、过程模式、价值观念、审美情趣的多方面变革。

我非常同意你的观点，建筑设计理论必须有建筑设计实践的深厚基础，实践的缺失是当前数字设计理论匮乏的根本原因。但是这些年来，也有不少好的理论书籍出版，如赖泽的著作《新建构图解》（2006年），以图解的方式对形体与物质、相似与相异等建筑问题进行了数字建筑语境下的阐释；博瑞的著作《脚本文化：建筑设计与编程》（2011年），对数字设计的常用方法与技术路线进行了研究；还有格博2014年的最新著作《计算的范式》探讨了数

and – for sure – we could say that the glory days of theory (that was so popular in the 1990s) are definitely over, even though – from my perspective – there is definitely something paradoxical about a 'death of theory' theory. But – for whatever the reason – it would appear that as yet we do not have any comprehensive theory of digital design.

The challenge, it would seem to me, is to develop a form of computational theory that emerges from within the domain of computation, and that could maybe even inform the way that we look at the analog world. The example that I like to quote in this regard is that of the 'boids' developed by Craig Reynolds as a way of modeling bird flocking behavior. In order to model that behavior Reynolds was forced to draw up some rules for it, and it was only after he had done so that biologists were able to understand the actual (analog) flocking behavior of birds. Incidentally, I got some sense of the potential of such a theory from a lecture given by John Frazer at the European Graduate School a few weeks ago. Here was one of the world's greatest living experts on digital design, who wrote the highly respected tome, An Evolutionary Architecture, and who collaborated with Gordon Pask and Cedric Price on the famous 'Fun Palace', beginning to espouse a theory of digital design. I was highly impressed by this.

XWG: The technical critic could help architect analyzing and elevating the design skills, but it has not reached the height of theory so far. It seems the time for theorizing digital architecture from architectural and technical history has not arrived yet. But I think philosophy of Deleuze is a good starting point for architects to establish their ideas and concepts of design, especially provide an abstract reference for the process of digital design. We should systematically investigate the theory, refine it into a fully developed theory for digital architecture.

On the other hand, complex science is the scientific foundation for digital architecture design. The territory of digital design could be greatly expanded if we apply the complex scientific theory and method on to the design process through algorithm modeling and coding. For example, we have had some applications currently in the complex adaptation system including Cellular Automaton, multi-agent system, Genetic Algorithm, Complex Network etc. But none-linear theories such as Fractal Theory, Chaos theory, Formal Syntax, and Artificial Neural Network have not applied much, and these will bring a bright future in the research. The research of complex science will lead to the revolution in the digital design method, design process, design value, design aesthetic and so on.

I agree with your opinion. Architecture design theory have to be based on an abundance of

字建筑设计的方法范式等。这些书的贡献在于对数字设计进行了归纳性阐述,现阶段还是特别需要这些书作为基础性理论的。

换个话题,您认为就数字设计而言,建筑实践所面临的最大挑战是什么?

NL:我认为西方建筑业已经深陷危机,而中国在过去几年中建筑业的发展势头很迅猛。建筑和建设问题与整体经济息息相关,如果经济环境不好,那么没有人愿意建造任何建筑,这将直接导致建筑师无事可做。与医生等职业不同,对于医生,社会上永远有一定量的需求,但是建筑师只有在经济繁荣的情况下才有更多需求。随着建筑师的主导性被蚕食,这个问题变得更加严重,越来越多的其他专业蚕食了建筑设计的工作。虽然各个国家的情况不尽相同,但是世界上平均只有大约5%的建筑是由建筑师设计的,因此可以说,建筑设计行业是萎缩了。而建筑教育也同样深陷危机。目前在西方,随着对建筑学有兴趣的学生逐渐意识到专业教育费用昂贵、耗时长且最终的经济回报也很少,建筑学专业的申请者数量每年以10%~20%的比例缩水,这些潜在的建筑系学生自然转去追寻更加可持续,并且有更多经济回报的专业,例如银行业。当然在中国这个情况非常不同,因为建筑行业仍然非常健康,但是在不远的将来可能会有所转变。中国的学术界目前发展不错,所以我们应当为未来做准备,并且保证建筑教育的适应性与生存能力,使西方所面临的问题不会在中国发生。

目前就数字设计领域来说情况并不太坏。我们甚至可以说建筑业面临着危机,而数字设计则能够提供一条危机之中的出路。因为数字革命所带来的是建筑业和其他专业之间边界的模糊。通过这些数字工具的使用,建筑师可以在多个设计领域一展才华。当各个专业的边界逐渐模糊之后,建筑师展现出多种可拓展的技术能力方面的天赋。例如,越来越多的建筑师如奈丽·奥斯曼、菲利普·比斯利、朱丽亚·克尔纳、弗兰西斯·贝多尼等,与像艾里斯·范·荷本这样的时装设计师合作设计三维打印的可穿戴物和其他的潮流物品。同样,我之前也提到在中国有马子聪三维打印的珠宝及可穿戴物和雷姆·D·库哈斯的鞋子。但是重点在于,时尚设计师本身并没有

architecture practice. The lack of practice is a fundamental cause for the lack of digital design theory. However during recent years, there are many good theory books emerging, such as "Atlas of Novel Tectonics" by Jesse Reiser (2006): by illustration of architecture issues as similarities & difference , form & material; "Scripting Cultures: Architectural Design and Programming " by Mark Burry (2011): the synthesis of common method and technical path for digital design; "Paradigms in Computing " by David Gerber (2014): the investigation into the sample of design method for digital architecture. Those books are of great achievements on the systematic introduction of digital design and especially needed as theoretical foundation in current era.

Now, let's move on to the next topic. What do you think about the biggest challenge for architectural practice in terms of digital design right now?

NL: Well, I believe that the architectural profession in the West is already in crisis. I know that in China the construction industry has been very healthy in recent years, but that is not the case in the West.The problem with architecture – and indeed with the construction industry as a whole – is that it is tied to the economy. If the economy is in a bad shape, and no one is proposing to build anything, there is simply no work for architects. Unlike professions such as medicine, where there is always a fairly constant demand for doctors, architects are only sought after in moment of economic prosperity. The situation has been exacerbated by the erosion of the architects' hegemony in the field. Increasingly other professionals are taking over in terms of designing buildings.While the situation varies from country to country, overall something like only 5% of buildings constructed around the world are actually designed by architects. As such, not only the profession of architecture but also the education of architects is in crisis. Indeed in the West we are seeing a reduction of between 10% and 20% in the number of applicants to study architecture, as increasingly would-be architects are realizing that they would be undertaking a long and expensive period of education with little financial remuneration at the end. Not surprisingly these potential students are now looking for more sustainable and economically rewarding professions, such as banking. Of course, the situation is very different in China – and the profession of architecture is still very healthy – but that might change in the not too distant future.Academics in China would do well, therefore, to plan ahead, and take measures to ensure the viability of an architectural education before the problem that has hit the West also hits China.

And yet the situation – in terms of digital design – is not so bad. We might even argue that

设计及生产这些物品的能力；事实上，建筑师的教育是非常丰富的，例如，建筑师具有三维思维和设计的能力，并且他们对于材料性能的理解远远超出了其他专业，只不过这部分具有极高可拓展能力的人误入了一个危机中的专业。在这种情况之下建筑师所要做的仅仅是拓宽自己的视野，并充分理解、运用建筑师所具有的能力；甚至可以说，AMO模式（一种空间顾问的形式）或许是建筑师的另外一条可走的道路。

徐：确实，数字革命带来了建筑业和其他专业之间边界的模糊，这使掌握了数字技术的建筑师很容易进入其他行业并获得工作机会。同时，具有数字技术才能的建筑师也有可能开创出新的领域，如以智能机器臂为工具，可开创自动化及智能化的建筑个性产品及加工制造行业；再如你上述提到的，以各种3D打印机为生产工具，可开创3D打印食品、3D打印玻璃制品、3D打印时尚穿戴物、3D打印建筑或其构件等。

但就建筑设计实践本身来说，数字设计不是孤立的行业，事实上它与数控加工及数字技术施工紧密相连，这三者形成了数字建筑产业链，它将是建筑产业升级的标志。在中国，它将有效解决劳动力成本上升、建筑加工及施工污染、从设计到施工各环节之间衔接不当而带来的浪费等种种问题；因而，将产生新的设计、加工、施工的组织方式。在数字设计的过程中，建筑师需要与其他专业的专家如结构工程师、软件工程师、材料工程师、加工厂商、施工技术人员等通力合作才能完成设计，而设计与加工及施工之间的联系方式将以数据及软件参数模型为媒介进行传递。建筑师对加工及施工的控制程度将极大地提高，建筑设计成果也将继续运用到建筑建成后的物业管理及建筑运营之中，这将进一步提高我们建造及管理居所的能力。从这一点来说，建筑师的职能将会改变，建筑师们应该从全产业链的角度重新思考自己的专业及其所应具备的技能，一个崭新的建筑师职业正在形成之中。

最后一个问题，您认为建筑教育领域所面临的最大挑战是什么？

NL：我认为目前建筑教育所面临的最大挑战是一些规范和管理的团体，例如NAAB美国建筑师协会以及他们对建筑的影响。目前

while the profession of architecture is in a potential crisis, digital design might even provide a way out of that crisis. Because what the digital revolution has brought about is an erosion of the distinction between architecture and other disciplines. With these new digital tools it is clear that architects can operate in many different fields. What is more this erosion of the boundaries of the profession has actually revealed how strong architects are in terms of their transferrable skills.For example, there has been an increasing number of architects – such as NeriOxman, Philip Beesley, Julia Koerner, Francis Bittonti and so on – who have been collaborating with fashion designers, such as Iris van Herpen, 3-D printing wearable and other fashion items.

Also, as I mentioned above, here in China, Steven Ma has also been a 3-D printing jewelry and wearable, and Rem D Koolhaas shoe. But my point is this: fashion designers themselves do not have the skills to design and fabricate these items. In fact the education of the architect is incredibly rich.or example, architects are able to think 3-dimensionally, to design, and to understand material behaviors in a way that no other professional is capable of doing. The situation we have then is that we have a sector of the community with highly transferable skills, who hitherto have been channeled into a largely unsustainable profession. As such, all architects have to do is to expand the understanding of what constitutes an architect. We might even say that the model of AMO – of a form of spatial consultancy – is another one that architects might pursue.

XWG: Indeed, digital revolution has blurred the boundary between architecture and other professions, thus architects with the digital technology would easily get access to other professions. At the same time, architects with digital skills could also find new territories, for example, the application of robotic arm could lead to the industry of customized and smart architecture. Also, as you have mentioned, production tools based on 3D printing technology could generate 3D printed food, 3D printed glassware, 3D printed fashion accessories, 3D printed architecture or components etc .

But for architecture practice itself, digital design is not an isolated field. In fact, it was deeply connected with CNC fabrication and digital construction. Those three forms an industry chain. It will be the sign of upgrade for building industry. In China, it will efficiently solve the problems of rising labor cost, pollution from construction, and the waste resulted from inappropriate connection between design to construction. Thus it will fundamentally changed the way organizing design, fabrication and construction. During the design process, the architect needs

看来没有一个类似的机构可以有效地保护建筑行业，因为由建筑师设计的建筑如此之少，而如果他们不能保护建筑行业，他们为何要如此严格地控制建筑教育？我感觉建筑系学生总有一天会反抗。事实上，我们可以看到一些像MIT这类院校的学生已经选择规划自己的教育，放弃NAAB认证的课程而选择一些未认证的课程，例如建筑研究，这样他们可以定制自我教育并且具有更自由的选课权；又如他们希望随尼尔·哥申菲尔德学习机器人制造或者选MIT媒体实验室的计算机课程，这些课程在NAAB认证的建筑课程系统中都是不可能的。进一步而言，这些地区性的组织极具地方保护性，但是在全球化的环境下基本没有意义，因为最好的建筑师都是在很多不同的国家实践。最夸张的是英国的建筑协会曾宣称伦佐·皮亚诺在英国不可以被称为建筑师，因为他在英国没有被认证，这有多么的荒谬呀！

更为具体而言，在数字设计教育领域面临的核心问题是我们缺乏足够多的具有数字设计领域博士学位的教授。在很多国家，包括中国，要求只有具有博士学位的人才能获得教职，而现在并没有足够多的大学提供数字设计的博士教育。这个原因也很简单，就是资源问题。数字设计是一个全新的领域，因此没有足够多的教授有博士学位并可以带学生。进一步讲，具有相关设施的学校也很少，就人力资源和设施条件而言，几乎没有学校可以提供这种服务。这是美国的现状，或许在其他国家更为严重，读博士是如此的昂贵，学生如果没有奖学金的话甚至无法读下去，在很多的美国大学仅仅学费便达到5万美金一年，还不包括生活费。考虑到大多数博士学位都需要7年左右来获取，而毕业后又会进入薪酬很低的行业，可能还助学贷款就要花费一辈子的时间。同时，建造业的赞助商对于研究的投入出奇的低，一部分原因是这个行业经济上的不稳定性；另一部分原因是建造业所建设的是一次性的建筑（每一座建筑都是独特的），与每种产品都被多次重复生产的手机和汽车行业相比，建造业更加不愿意在研究中投入。结果导致在数字设计领域中没有足够多的博士毕业生来满足学术界的需求，这也是急需解决的一个问题。

to cross discipline to communicate with the experts such as the structure engineer, software engineer, material engineer, fabrication company, construction technician to complete design. The communication between design, manufacture and construction will be based on data and digital model, thus providing the architect with greater ability to control the level of fabrication and construction. Even some design products such as BIM model will continue to be used for the maintenance and management of the building. In this aspect, the architecture profession will change. Architect should rethink their profession and skills from the aspect of full industry chain. A brand new architecture profession is at the verge of emerging.

And what do you think about the biggest challenge to architectural education right now?

NL: To my mind one of the biggest challenge facing education today is the stranglehold that certain regulatory bodies such as NAAB another national registration councils hold on architectural education. It seems that none of these bodies are very effective in protecting the profession of architecture, in that only so few buildings are actually designed by architects. If they are unable to protect the profession of the architect, why do they have such a stranglehold over architectural education? I sense that sooner or later students are going to rebel against this. Indeed we can even see the example of students in schools such as MIT where they are taking charge of the their own education and opting out of the NAAB accredited Architecture degree and choosing instead degrees in non-accredited courses, such as Architectural Studies, where they have the freedom to customize their education and choose courses with a greater degree of freedom than the very limited and constrained NAAB accredited program. For example, they might wish to study robotic fabrication with Neil Gershenfeld, or some course on computation at the MIT Media Lab, which wouldn't be possibly under the much more constrainedNAAB accredited program in Architecture.What's more these local regulatory bodies are highly provincial and have become very territorial, and yet they make little sense in a world that has become thoroughly international and where many leading architects are practicing in several different countries. The most absurd story that I heard recently was that the registration council in the United Kingdom decreed that Renzo Piano could not be called an architect in the United Kingdom because he was not qualified in that country. How absurd is that?

But in terms of the more specific question of digital design education the key issue is the shortage of professorswith PhDs in digital design right now. In many countries – including China – it is becoming an absolute necessity to have a PhD in order to be offered an academic

在西方,我们最近创立了欧洲研究生院名下的一个非常有趣的项目。在这里我们创立了一个数字设计的博士学位,并且吸引了一些世界上最好的数字设计的先锋代表来授课,包括数字设计的教父 J.约翰·弗雷泽、阿希姆·门格斯、帕特里克·舒马赫、马克·贝瑞、凯西·瑞斯、阿丽莎·安底拉斯克、弗朗索瓦·罗氏,以及菲利普·比斯利等,构成了全世界最顶尖的教授名录。就算 MIT 也不能与其相媲美。同时,它还是一个在职的课程,这样学生只要每年参加三周的密集课程,其他时间可以在世界的其他地方持续实践或者做学术性的工作。这个项目与大多数美国的项目相比不仅仅就学费而言进入了大众可承受的范围,同时它也使得学生在就读的同时保证了日常的收入来源,使得整体费用相对便宜。这对于博士教育而言是一个巨大的变革。在中国也有类似的事情正在发生,事实上我在同济大学所扮演的一个重要角色便是作为中国政府所支持的"高级外国专家"聘请计划的访问教授,在同济大学建立一个和欧洲研究生院类似的博士项目,不过这个项目更多关注实际的生产和建造,因为中国目前不仅仅是世界工厂,更是正在走向世界前列的数字加工中心。简而言之,在中国我们需要有更多的数字设计的博士生来为学术界服务。

徐:从宏观角度来说,无论政治、经济、社会,还是科学、技术,以及建筑实践,都已经产生了新的需求,而目前的建筑教育已经不能满足要求,我们面临的最大挑战在于如何寻求新的建筑教育模式以培养新一代建筑人才。作为主讲人,我多次参加了 IAES 国际建筑教育会议,2009 年东京会议提出"基于全球化的建筑实践培养全球化的建筑师"以及"培养跨界的建筑师";2013 年柏林会议讨论了"传统院校之外的建筑教育平台"、"建筑教育如何实现专业交叉"、"建筑教育如何与非学术机构合作"等议题;2015 年新加坡会议以"建筑教育正在涌现的网络化"为主题,着重讨论了"建筑院校间的联合"、"技术对建筑教育的影响"等问题。我认为这些问题的讨论有助于目前建筑教育的改革,并可以促进世界各国建筑院校教育模式的创新。参加会议的主讲人均来自世界领先的建筑院校以及著名的建筑事务所,或者来自于政府机构,

appointment. And yet there are simply not enough people studying for a PhD indigital design at universities. The reason for this is simple. It is a question of resources. Digital design is such a new field that there are very few professors with a PhD who can themselves teach new PhD students. Furthermore, there are relatively few institutions with the right facilities. In other words, in terms of both human and technical resources, there is almost no institution capable of serving this function.The situation in the States is perhaps even worse than in other countries, in that it is so expensive to study for a PhD that students cannot even start until they have some form of scholarship. In many US universities it can cost up to $50,000 a year to study – simply in terms of fees, and not including living expenses. If we add to this the fact that most architectural graduates will have studied for 7 years just to obtain their masters degree with the prospect of going into an underpaid profession where it might even take a lifetime to pay back student loans, the situation is quite serious.Moreover, there are few prospects of sponsorship in that the construction industry is notoriously underfunded when it comes to research.This is partly because of the economic instability within that industry, but partly also because the construction industry is focused primarily on one-off buildings – each of which is different. Compared to the mobile phone industry, or automobile industry, for example, where each product is reproduced many times, the construction industry is ill equipped to invest in research. The result is that there are insufficient PhD students graduating in the field of digital design to service the needs of academia. Something needs to be done about this problem.

In the West we have recently launched a very interesting initiative as part of the European Graduate School, where we have set up a PhD program in Digital Design, and have managed to attract some of the world's leading voices in digital design to teach on it. Figures, such as John Frazer – one of the 'godfathers' of digital design – AchimMenges, Patrik Schumacher, Mark Burry, Casey Reas, Alisa Andrasek, Francois Roche and Philip Beesley, are all teaching on this program, giving it the greatest range of the leading experts anywhere in the world. Not even MIT could compete with this. Also it operates as a part-time program, so that students only have to come for 3 intense weeks of lectures once a year, meaning that they can maintain their jobs elsewhere – whether they are working in practice or academia. Not only is the program much more affordable in terms of fees than any program in the States, but it allows students to maintain their regular source of income thereby making the overall package relatively cheap. This is a major revolution in terms of PhD education. But something like this also needs to happen in China. In fact one of

他们可以直接把讨论的结果带到当地的建筑教育实践中。同样，这些讨论的议题也可以作为任何其他建筑院校思考的问题，通过讨论和思考，建筑教育应该能找到满足新时代要求的，并且具有特色的、适合各自特点的新的教育道路。

确实，在世界建筑教育探索新方向的大背景下，建筑教育作为基础，为中国数字工厂的发展起到支撑的作用，除了你刚才提到的DADA每两年一次的系列活动所展现出的成果外，每年还有一次全国建筑数字技术教学研讨会，今年的会议出现了许多具有示范性的教学成果及研究成果，比如清华大学建筑学院已形成从本科低年级到研究生的系列数字设计教学课程体系，包括基本的软件学习、设计思想及方法的教学、专门的数字设计 studio，以及机器臂数控建造的实验性教学，这在世界上可能也是屈指可数的范例；清华的毕业生受到业界的欢迎；再比如同济大学建筑与城市规划学院建立了令人羡慕的设备齐全的数字建造实验室，可进行各种教学及科研的数字建造实验。东南大学"建筑运算及应用实验室"不仅拥有CAAD、GIS、VR技术支撑的各类CNC数控制作和实验平台，包括KUKA机械手臂、折弯机、三维打印机，并且具有配合建筑互动设计实验所需的电子检测设备及印刷电路板制作系统，为研究及教学提供了有效的平台。中国多数建筑院校过去都设有CAAD教学课程及教师，现在几乎都转型到关注数字设计的方向上来了，这些教育上的转变将有效推动数字工厂的发展。

my roles at Tongji University – where I have recently taken up a Visiting Professorship funded by the Chinese Government 'High End Foreign Expert' recruitment initiative – is to set up a PhD program broadly along the lines of the European Graduate School, but focusing more on actual fabrication and building construction, as China is rapidly turning into not only the Factory of the World but also the leading Digital Fabrication center in the world. We simply need to see more PhD graduates in Digital Design in China in order to service academia.

XWG: From macroscopic point of view, new needs have emerged from politic, economy, society, and science, technology, architecture practice as well. Currently our architecture education can not satisfy the need anymore. The greatest challenge we are facing is how to seek for new format of education and educate the next generation of architects. As a speaker, I have attended the International Architectural Education Summit multiple times. During the 2009 Tokyo summit, we raised the issue of "architecture practice based on the global practice and the education of global architect" and the "education of the architect beyond boundaries". During the 2013 Berlin Summit, there is discussion about "The role of alternative architecture education platforms", "Interdisciplinary strategies in architecture education", "Collaboration between architecture education and non-academic partners" etc. During the 2015 Singapore conference, the topic includes "Emerging networks in architectural education "focusing on the "alliances between architecture schools", "the influence of technology on architecture education" etc. I feel discussion of these issues would help with the reform of architecture education and catalyze innovation of education system around the world. The speakers of the summit are from top architecture schools, prestige architecture firms around the world, and from governments. They can use the result of the summit directly to influence local education. At the same time, architecture schools should think about those topic as well. Through thinking and discussing, architecture educations will find distinguishing paths that is tailored towards their own realities confronted in new era.

Yes, under the global trend of searching for new direction for architectural education, Chinese architectural education seems to have found the direction on digital architectural education. Architectural education as the base supports the development of 'Digital Factory' in China. Beside the achievements of the biennial DADA event series as you have mentioned, there is also a yearly national education seminar on digital architecture technology. This year, many provocative samples have presented from the teaching and academic researches. For example, there are a systematic course structure on digital design in Tsinghua Architectural

Department, spanning from undergraduate to graduate educations, including software learning, design strategy, digital design studio, and the experimental studio for robotic construction. This is a rare example globally. The graduate of Tsinghua Architecture is welcomed by many prestige architecture institutes around the world. Also, Tongji School of Architecture and Urban Planning has put together a digital fabrication lab with an admirable collection of equipment. The "Digital Architecture and Application Lab" in Southeastern University not only has the CNC experiment platform supported by CAAD,GIS,VR including KUKA robotics, folding machine, 3D printer, but also has all the necessary electrical evaluation equipment and circuit board testing system, providing a efficient platform for research and education. There were CAAD course and teachers in many Chinese architecture schools, now nearly all of them have shift their focus onto digital design.

美国加州艺术学院 / CCA

几何编织者 /
Geoweaver

几何编织者是一个大型可移动机器人3D打印平台的原型。它采用先进的机器人,有能力来扫描并建立周边环境的模型,并最终在原地打印出整个地形及房屋。打印机可以通过无人机部署,可以同时在极端地形中穿梭并进行混凝土打印作业。几何编织者通过最新的附加3D技术,融合了"六足虫"的体形以及机动性。该项目作为加州艺术学院(CCA)数字工艺实验室(Digital Craft Lab)的创造性建筑机器工作营(Creative Architecture Machines studio)的一部分,运行了两个月,并使用Arduino,Grasshopper以及Firefly作为主要的硬件及软件平台。

Geoweaver is a prototype for a large-scale mobile robotic 3D printing platform. It utilizes advanced machine vision protocols to scan and model its environment, and ultimately print entire land forms and buildings in situ. The printer, deployed via a drone, can simultaneously traverse and print concrete in extreme landscapes. The Geoweaver project hybridizes the body-type and mobility of hexapod insects, with emerging additive 3D technologies. It was developed over two months as a part of the Creative Architecture Machines studio at the CCA Digital Craft Lab in San Francisco, and utilizes Arduino, Grasshopper and Firefly as its primary hardware and software platforms.

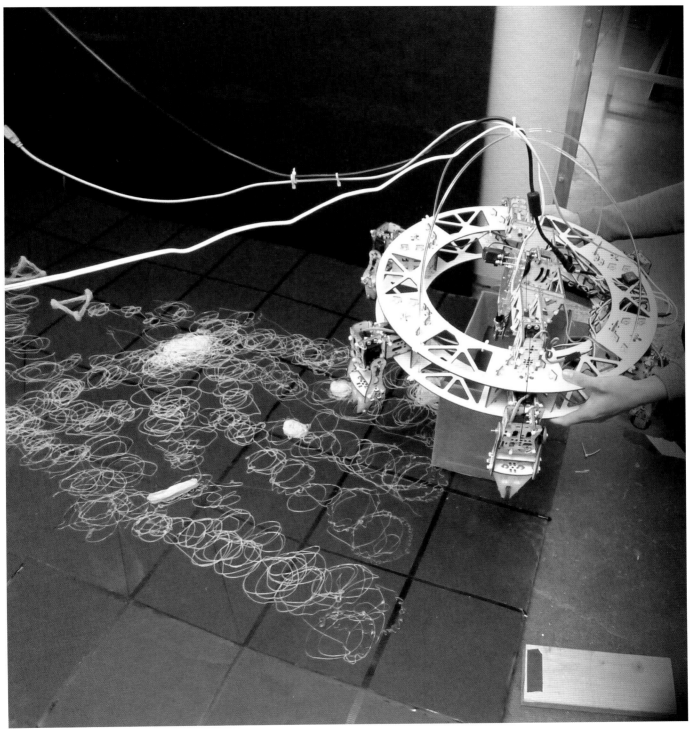

层状网络系统 /
Stratum Networks

层状网络系统是一个大型的定制化"实时"三臂机器人打印平台的原型。它是一个实时的 3D 打印机——它的引导机器代码可以进行实时调整以实现不同的效果。它也可以与传感器及其他实时输入设备配合使用，从而将独特的场地环境或者声音和使用者的反应等不可预测的变量结合起来。层状网络系统将传统的陶瓷艺术和黏土挤压技术，与最新的三臂机器人的体形结合起来。该项目作为加州艺术学院数字工艺实验室的创造性建筑机器工作营的一部分，运行了两个月，并使用 Arduino，Grasshopper 以及 Firefly 作为主要的硬件及软件平台。

Stratum Networks is a prototype for custom "real-time" deltabot type large-scale printing platform. It is a live 3D printer –its guiding machine code can be tweaked in real-time to produce a variety of effects. It can also be coupled with sensors or other live inputs to incorporate site-specific or unpredictable variables like sound or user interactions. The Stratum Networks project hybridizes the body-type of novel three-armed deltabot, with more traditional ceramic arts and clay extrusion technologies. It was developed over two months as a part of the Creative Architecture Machines studio at the CCA Digital Craft Lab in San Francisco, and utilizes Arduino, Grasshopper and Firefly as its primary hardware and software platforms.

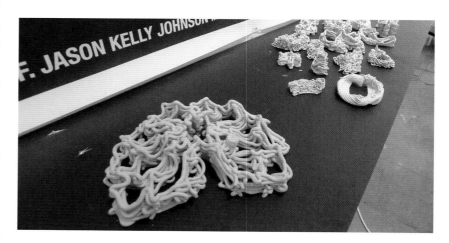

集群式造景器 /
Swarmscapers

集群式造景器是一个可完全自主移动的 3D 打印平台的原型。它利用了先进的机器视能和集群算法，以指导自身穿越周边环境，并在锯屑的表层实施紧急的三维打印作业。集群式造景器结合了军用坦克或者越野车的形体，以及在传统的淀粉打印技术基础上建立起来的分层挤出技术。该项目作为加州艺术学院数字工艺实验室的创造性建筑机器工作营的一部分，运行了两个月，并使用 Arduino，Grasshopper 以及 Firefly 作为主要的硬件及软件平台。

Swarmscapers is a prototype for a fully autonomous mobile 3D printing platform. It utilizes advanced machine vision protocols and swarming algorithms to guide itself through its environment and print emergent three-dimensional forms in layers of sawdust. The Swarmscapers project hybridizes the body-type of a military tank or rover, with layered extrusion technologies found in common starch printing technologies. It was developed over two months as a part of the Creative Architecture Machines studio at the CCA Digital Craft Lab in San Francisco, and utilizes Arduino, Grasshopper and Firefly as its primary hardware and software platforms.

美国哥伦比亚大学 /
Columbia GSAPP

满足中心 /
Fulfillment Center

使用 Processing 软件模拟移动建筑构件的物理性动态，这个"满足中心"汇集了亚马逊网的逻辑算法和日本情人旅馆的享乐。物理模拟及概率建模的工程师用精确但混合着经验的方式不断地重组着这些满足要求的组合方式。偶然的相遇是基于模拟一个连续搅拌的微型空间之上，为附近的秋叶原地区的需求提供服务。

Using Processing to simulate the physics-based dynamics of moving building components, this "fulfillment center" brings together Amazon.com logistics algorithms and the hedonistic pleasures of a Japanese love hotel. Physics simulation and probabilistic modeling engineer precise but always changing mixtures of experiences to continually reform this all-satisfying array. Chance encounters are built into the simulation through a continuous agitation of micro-envelopes, delivered on-demand to the neighborhood of Akihabara.

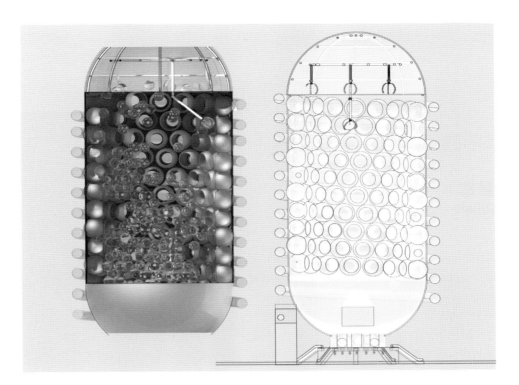
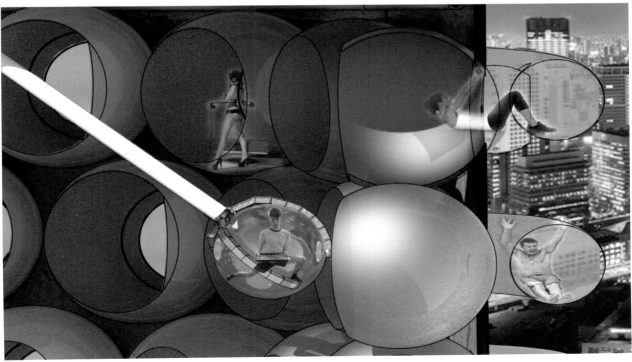

像素 /
Pixels

该项目采用数字仿真塑造并充斥着东京摩天大楼的巨大立面，立面上的智能模块吸收太阳辐射，通过铺天盖地的像素化广告牌表达自己。每个在 Processing 中被计算、塑造的智能模块，通过权衡它们的位置、周边立面可见性以及日照潜力，智能地在全球能源生产、广告收入和日常活动中获得平衡。当它们聚集时，就形成在消费和生产之间平衡交替的建筑分布。

The project uses digital simulation to model and occupy the immense facades of Tokyo's skyscrapers, where intelligent drones absorb solar radiation and express themselves as pixels in an omnipresent billboard architecture. Each drone, modeled as computational agents in Processing, weighs its position, the visibility of nearby façades and their solar potential to reach intelligent compromises between global energy production, advertising revenue and daily movement. Together, they form a radically distributed architecture that alternates between modes of consumption and production.

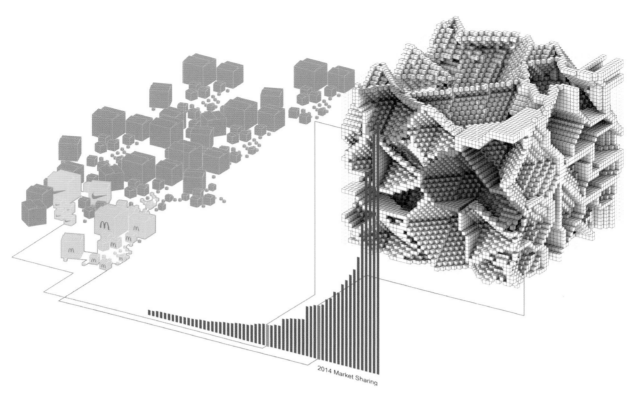
2014 Market Sharing

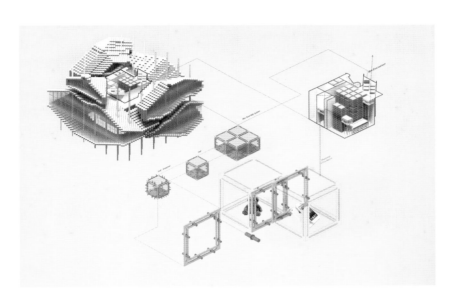

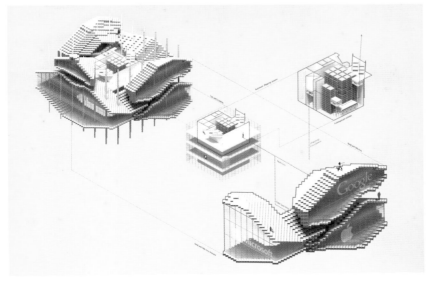

临时性空间 /
Provisional Spaces

城市内部空间随着东京秋叶原电子产品区域物流中心的转移呈现扩张和破裂。每处空间内的组织和分裂通过在 Processing 中的适应性增长、塑造被加以算法式地生成。利用历史财务数据作为依据，每个单元模拟的是美国和日本高科技公司的成长模式，其结果是一个临时的空间网络，这些空间以起伏的供需关系进行扩张和破裂。

Interior envelopes expand and collapse with the shifting inventory of an urban fulfillment center in Tokyo's Akihabara electronics district. The organization and divisions between each space are generated algorithmically through adaptive growth, modeled in Processing as a set of competing agents. Using historical financial data as a training set in disruption, each agent mimics the growth patterns and affiliations of a US or Japanese tech company. The result is a network of provisional spaces that opportunistically expand into vacancies and intertwine with the ups and downs of supply and demand.

美国哈佛大学设计研究生院 /Harvard GSD

光之森林 /
Light Forest

该互动装置试图重塑一种与他人产生关联的移动方式。为了实现这一目的,一个螺旋色带的阵列被建立起来,这个色带阵列捕捉人的出现并对其做出反应。螺旋色带连接着快速旋转的马达,这些马达感应到驻波,并以马达转速的一定比例使色带收缩。当一个人出现在阵列中,在她面前的色带将会收缩,从而使她通过。当两个人出现时,该系统将会以不同的方式运作。在一条连接两人的路径上的丝带将会打开,以驱使两人在某一点上相遇,从而增加了社交的可能性。

In this interactive installation there was an attempt to reshape the way people move in relation to each other. To achieve this an array of spinning ribbons was created that tracks and responds to the presence of people. The ribbons are attached to fast spinning motors that induce standing waves causing them to contract in proportion to the motor speed. When a single person is present the ribbons in front of her contract to enable her to pass through. When two people are present the system works differently. The ribbons that lie along a path connecting the two people open up forcing the two to meet at a point and thus increasing the likelihood of social interactions.

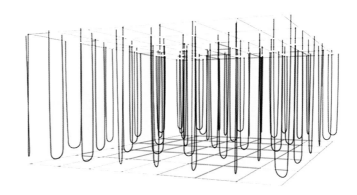

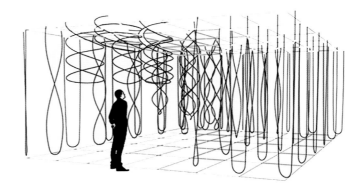

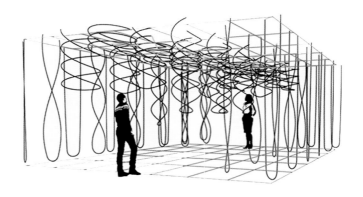

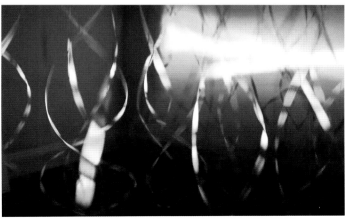

建造原型的（再）建构 /
(Re)Fabricating Nervi

皮埃尔·奈尔维通过对于预制构件的实验，以及他对混凝土结构优化原理的直观理解，发展出了一系列专利。这一策略向世人提供了一个具有鲜明审美情趣的高效系统。该项目以此为基础，经由对几何变体以及交变应力模式的系统性探索，引入了新的材料以扩大奈尔维的假设领域。该项目试图充分利用新的建构过程以扩展奈尔维的基本原理，同时避免了由于该项目的复杂性所带来的高强度劳动的施工过程。该项目也探索了结构的主要受力线的优化，同时还探索了多种元素组合的高效的建构与装配方法。

Pier Luigi Nervi developed a series of patents through the experimentation with precast elements informed by an intuitive understanding of principles for optimizing concrete structures. This strategy provides for an efficient system with a distinct aesthetic sensibility. Building on this via systematic exploration of new geometrical variations and alternate stress patterns, this project introduced new materials to expand the scope of Nervi's hypothesis. It sought to take full advantage of new fabrication processes in order to expand on Nervi's foundational principles but to remove the labor intensive construction process that the complexity of the projects necessitated. It also explored the optimization of a structure's main stress lines in parallel with efficient fabrication and assembly methods of multi-element aggregations.

通感展亭 /
Synaesthetic Pavilion

通感是一种通过直接刺激,造成一种感觉通路向另一种感觉通路转变的无意识的认知反应的触发。该展览馆是通过一个基于克拉尼实验的算法(这种算法里声波将朝向振动率为零的波节线振动),提取声音中的信息并将其转变为一种可视化的组织方式而生成的。不同的声波振动以不同的方式重新组织物质以及几何形体。几何形体的复杂性随着声波振动频率的升高而增加。该项目将克拉尼法则发展到一个三维几何形体探索的领域,以实现展览馆的物质性建构。

Synaesthesia involves the involuntary triggering of a cognitive response from one sensory pathway through the direct stimulation of another. The pavilion is generated through the transposition of information extracted from sound into a visualized emergent material organization, through an algorithm based on the Chladni Experiment in which sound wave vibrate towards the nodal line where vibration is zero. Different sounds vibrate and reorganize matter and geometry in a different way. The geometrical complexity increases with the increase in sound frequency. This project extends the Chladni Law into a three dimensional geometrical exploration that allows for the material configuration of the pavilion to emerge.

美国麻省理工学院 /MIT

编织的操作 /
Knitting Behavior

编织行为是一种以物质为中心的设计方法,用来探究织物的纤维构成是否能产生一类材料特性,它可延展而用于建筑形式的设计生成。这个项目涵盖了物理建造、可比的结构荷载以及数字模拟,从而探索材料特性,如硬度、弹性以及方向性,最终在数字模型中提供材料的性能。基于此研究,曾有的编织模式能够具有动态价值,并且从数字及物质两方面生成图形,以适用于建筑形式和行为需求。

Knitting Behavior is a material-centric design methodology which tests whether the composition of fibers in a knitted textile can produce a range of material attributes useful for the design and generation of tensile architectural forms. The project employs physical making, comparative structural loading, and informed digital simulations to explore the materials' active characteristics - such as stiffness, elasticity, and directionality – in order to implement true material properties into a digital model. With this, traditionally inactive knit patterns can incorporate dynamic values, allowing patterns to be digitally and physically generated to suit the behavioral needs of tensile architectural forms.

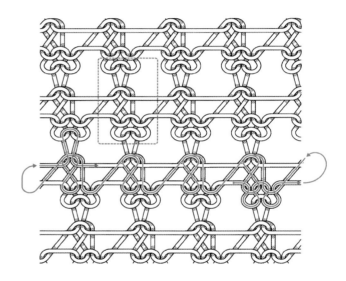

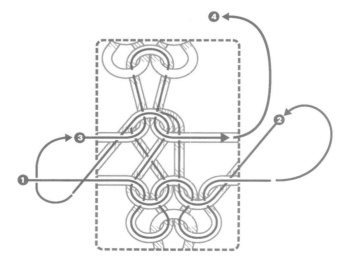

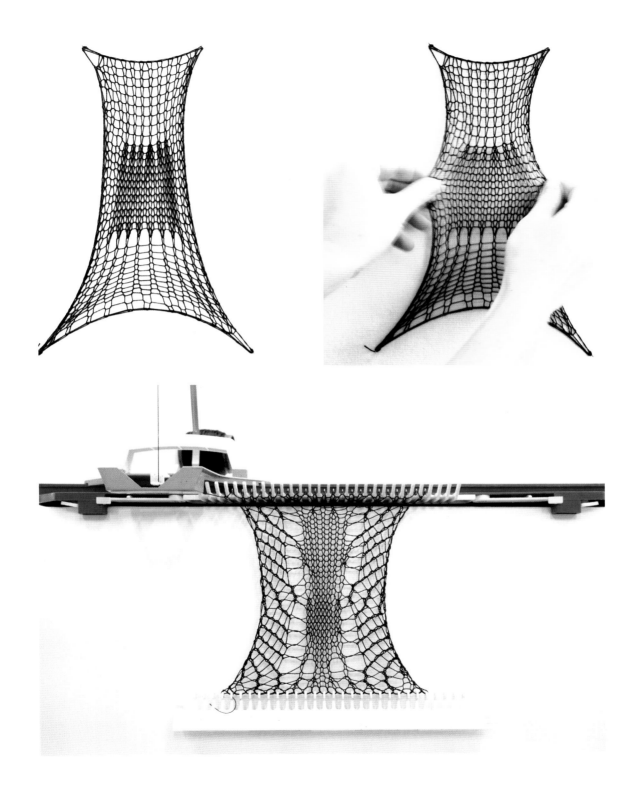

做手势 /
Making Gestures

"做手势"提出了一个互动模型,其试图超越如今建筑设计实践中不可避免的身心二元,以实现更具互动的计算制造形式。探究思维、身体以及设计工具的实时互动,通过身体的语言与建造机器的输入,实现人工智能的对话,其包含的模糊与意外使建筑师更加投入于精深的设计和建造过程中。

'Making Gestures' proposes a model of interaction that seeks to transcend the hylomorphic model imperative in today's architectural design practice to a more reciprocal form of computational making. To do so, it explores real-time interaction between mind, body, and design tools by using body gestures and imbuing fabrication machines with behavior using artificial intelligence in order to establish a dialog which embraces ambiguity and the unexpected to engage architects into insightful design and fabrication processes.

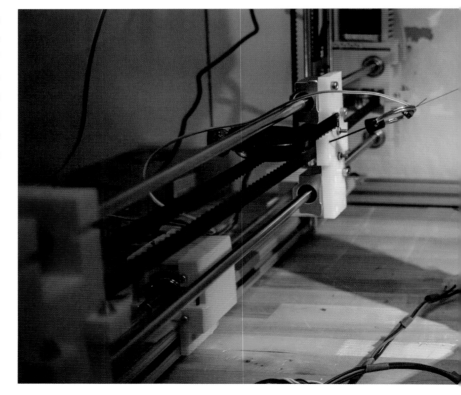

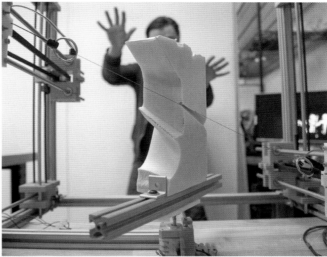

向大师学习 /
Learning From the Master

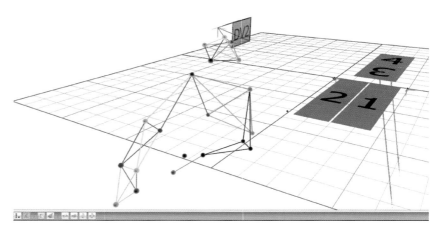

本项目设计并制作了一个能在新手学习程序技能过程中提供帮助的可穿戴设备，目标是减少熟悉新技术和媒介所需的时间，并迅速达到基本的水平。通过对比学生正在进行的与提前录制好的大师的动作，为初学者提供连续视觉反馈。通过放置在学生前臂上的发光二极管以及记录运动的加速度计，磁力计，陀螺仪和肌肉电活动仪，能实时利用肌电图进行记录。

The objective of this project was to design and implement a wearable device that aids novices during the skill acquisition process of any procedural motor task. The goal of the wearable device is to significantly reduce the amount of time needed to familiarize oneself with a new technique and medium, and to quickly attain a basic level of proficiency. This is achieved by providing students continuous visual feedback, which compares their on-going movements to that of a master craftsman performing the identical task recorded beforehand. Illuminated LEDs placed on the student's forearm relate movement kinematics, an accelerometer, magnetometer, gyroscope and muscle activity, all of which are recorded using electromyography (EMG) electrodes in real-time.

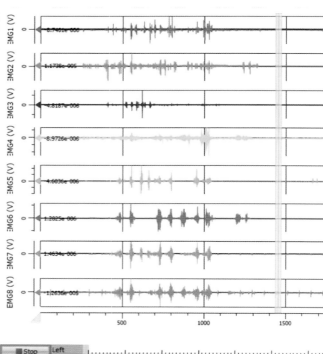

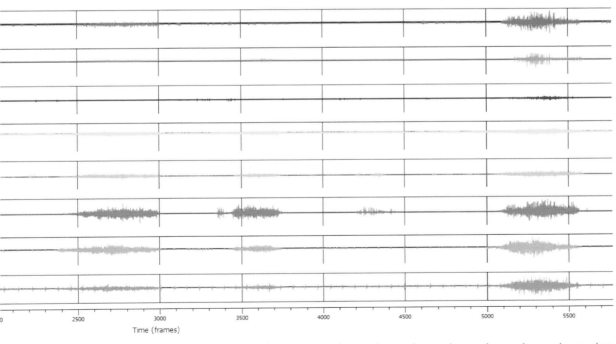

Time (frames)

美国普瑞特艺术学院 /Pratt

模糊线 /
Blurred Lines

模糊的线条是在新材料制备过程中涉及控制阈值的塑造过程，这一过程是输入计算机代码和输出物理纤维共同作用的结果。我们建造了熔融沉积式建模的三维打印机，取消打印的部分支撑物，所得的结果是"暴跌"线或链。我们研究介于液体和固体塑料之间的"玻璃"的状态，这会获得意外效果。塑料的黏度、挤压速度、挤压方式、室内温度和重力因素都对计算机数字和物理物质间的"保真度损失"产生影响，生成非凡的幻影形式。

Blurred Lines is a project about the modulation of the threshold of control in a new material fabrication process that requires a digital input of lines and generates a physical output of fibrous interactions and weaves. A series of Fusion Deposition Modeling 3D Printers were built and then hacked. The result of this process was the "slumped" line or the catenary. The viscosity of plastic, as well as the speed at which it was being extruded, the pattern of extrusion, the temperature of the room, and gravity contributed to a "loss of fidelity" between the states of digital and physical matter. This loss generated a physical language where the invisible became ghosted within extraordinary forms.

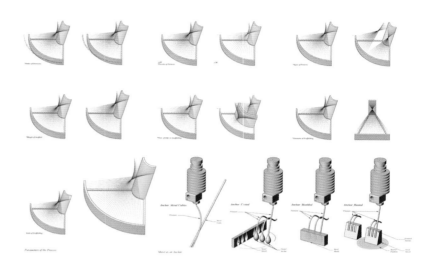

雪天使 /
Snow Angel

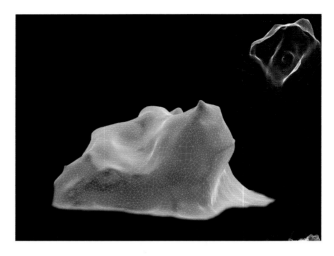

雪天使是一个世俗的建筑，它是为了庆祝生命的最后时刻。在半低温和半高温的状态下，冻结的组织会产生热交换。空间之间的体积关系是源于对巴洛克风格教堂的分析，其几何形式是基于宗教中生与死的纠结。在这一人工冰山中，我们使用了两个模拟的系统去塑造建筑原型。在细部层级，我们把黑色丙烯酸涂料和铁屑加以混合，雕刻出柱廊纹路，之后进行三维扫描。在空间层级上，我们利用地形建模算法生成霜化／熔化的表面与毛刺，并通过 3D 打印进行进一步观察。物理模型并不仅是一个物理和数字模拟的结合，而是一种批判性的评估方法。

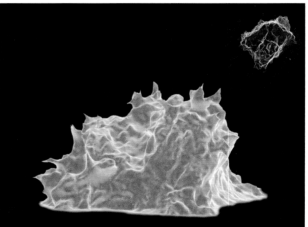

Snow Angel is a secular architecture for the celebration of the last moments of life. Half cryogenic cemetery and half hot spring resort, the freezing of tissue causes heat exchanges that boil water. The volumetric relationships between spaces are derived from an analysis of baroque cathedrals, where geometries entwine the worshipping living with the worshipped dead. In this artificial iceberg, the project used two analogous systems. At a detail level, it mixed ferrous acrylic paint with iron filings, and sculpted pools into columns, and 3D scanned them. At a spatial level, it leveraged terrain modeling algorithms to generate frosted surfaces, which were further tested with 3D prints.

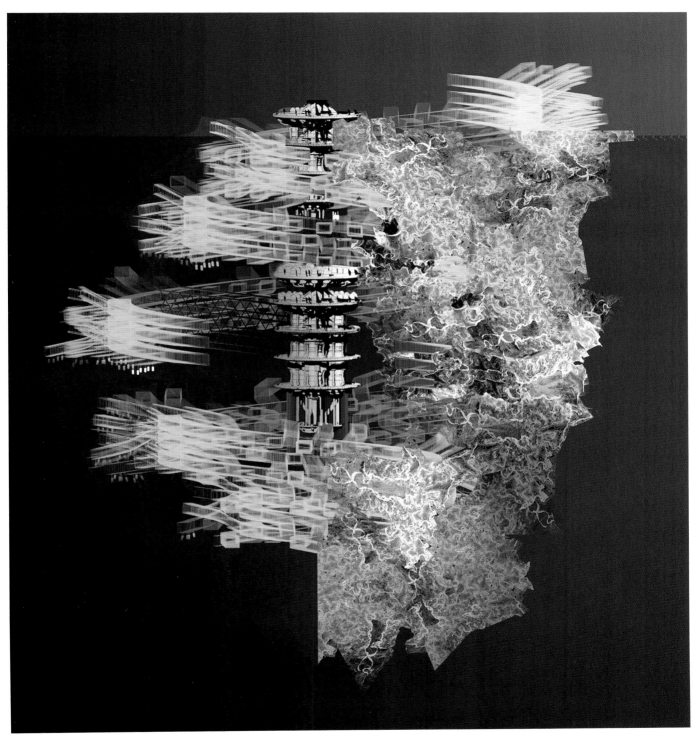

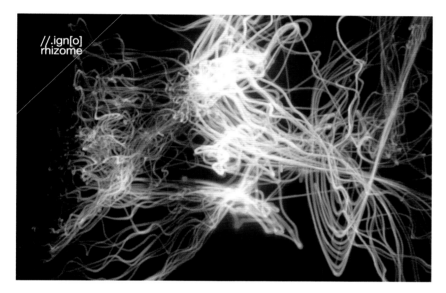

块茎 /
Ign[o]rhizome

研究的目的是表达一个多代理系统及非典型的行为，以此创建一个高度易变的绘图机。在系统机构的基础上，对其他代理行为进行编码以创建局部异常，这可以生成一个新的全局秩序和组织形式。标注行为的时间是另一个方面的研究，利用色彩表现时间（世代）以及局部密度（代理交互）。研究的另一个目的是维持辩证算法和机器人之间的关系，以某种方式绘制真实的三维空间，而非模拟。

The intent of the research was to impart a multi-agent system with non-familiar behaviors in order to create a highly volatile drawing machine. While stigmergic behaviors were the basis for the agency of the system, other agent behaviors were encoded in order to create local anomalies which could contribute to a new global order and composition. Annotating behavior through time was another agenda of the research, utilizing color to register time (generational) as well as local density (agent interaction). Another intent of the research was to maintain a dialectic relationship between algorithm and robot in order to somehow materialize or draw in true 3D space, rather than to simulate it.

美国普林斯顿大学建筑学院 /Princeton

光之媒介 /
Light as Creative Medium

本方案是一个关于由一个房间尺度内反射物体产生的复杂光线下交互过程的实验。反射物体将一个单一透射体发出的光线散落在一个封闭空间中的墙体上，形成了复杂的图样。空间的使用者或者设计者可以直观地利用伺服系统或者红外线，并通过旋转反射物体对墙上的光纹进行操纵。软件绘制了二维投射图与设计者手臂的三维坐标及在伺服系统上的旋转角度。通过这样的绘制，设计者可以对光线的结果进行控制以及对其位置精准定位。

This project is for an interactive process of experimenting with complex light effects produced by reflective objects at the scale of a room. The reflective object scatters the light of a single projector into a complex pattern on all walls of the enclosure. The user/designer of the space is able to intuitively manipulate the scattered light by rotating the reflective object by means of an Arduino-powered servo and infrared hand-tracking with a Kinect. The software maps the relationship between a two-dimensional projected image and the servo's angle of rotation onto the three-dimensional coordinates of the designer's hand, and it is through this mapping that the designer can exercise some control and precision over the location of the unpredictable light effects.

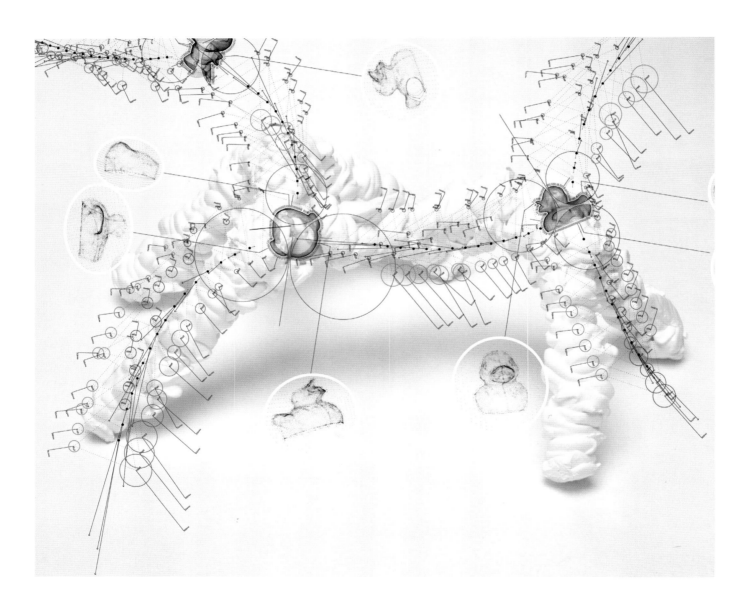

制造呈现 /
Reimaging Fabrication

用于制造的工具已经迅速地发生变化，而对建造时间的绘制和图像化却仍然保持不变。制造需要一种新的具有代表性的语汇来与现实进行连接。"制造呈现"塑造了一种合成的具有代表性的环境。这种重构技能是仪器直觉与材料偶然的交界。"制造呈现"通过对材料进行计算和供人分享的仪器化方式重新创造了一种建筑的非假想的形式。

While the tools for making have rapidly changed, the translation of construction practices into drawing has remained relatively intact. Fabrication requires a new representational vocabulary in which translation is imaged in contact with the real. Reimaging Fabrication develops a synthetic representational environment. This reskilling is an interface between instrumental intuition and material contingencies. Reimaging Fabrication promotes an architecture of non-ideal form through material computation and participatory instrumentation.

坦塔罗斯的傲慢 /
Tantalus' Hubris

随着一定程度智能性和适应性的植入，机器人也许不仅仅可以执行预先设定的工作程序去制造预先设计的形式，也能够根据变化的刺激因子去产生动态的、交互变化的形式。类似 3D 打印机，机器人用根据预先设置的工作程序来沉淀可融化的材料。然而，机器的运转速度可以根据不停地反馈来改变，这个反馈来自于目标交互的过程及镜头监测到的实施沉淀情况。这样，机器人变成了一种现实中的可即兴编辑设计的交互媒介。

With a certain embedded intelligence and adaptability, robots may be used not only to execute predetermined tool paths to fabricate predesigned forms, but also to react to changing stimuli to generate dynamic, interactive forms. Working like a 3D printer, the robot is used here to deposit molten material along a predefined tool path. However, the speed at which the robot moves from point to point is determined by a responsive feedback loop between by an interactive simulation defining an editable target path and a camera detecting the progression of real deposition. In this way, the robot becomes an interactive agent in the realization of an improvised and editable design scheme.

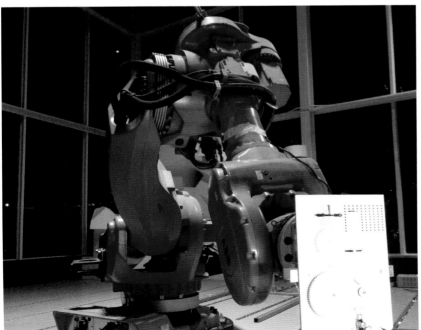

美国伦斯勒理工大学 /RPI

混合方法论 /
Methodological Hybrids

工作室通过两种不同程序生成的形式,探讨一种混合的设计观点。周期性的几何镶嵌和复杂算法生成的形式作为两个不同的系统被引入,以彰显不同区域住宅项目的潜力。周期性网格表现重复的较小规模的私人空间,复杂算法则生成表现较大规模公共区域的空间和结构。这两个系统的交叉部分创造了新形式,包含新的潜在功能和项目,把严格的周期性空间变换为流动的公共空间。

This housing studio explored a hybrid design mythology through two different procedurally motivated methods of generating form. Periodic geometric tessellations and complex algorithmic methods for generating from were introduced as two different systems of formal arrangement, each with the potential of informing different zones of the housing project. Manipulations of periodic grids informed the arrangement of repetitive smaller-scaled private spaces; algorithmic methods informed the space and structure of larger-scaled public areas. The intersection points of these two systems were mined for formal innovations that contained potential to hybridize new functions and programs, unraveling harsh periodic spatial repetition into fluid public spatial configurations.

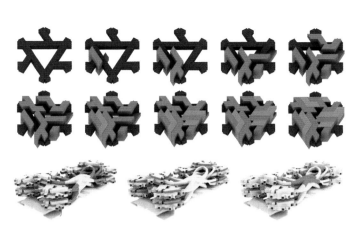

像素化有机体 /
Pixilated Organicism

在纽约特洛伊住房项目中，建筑的形式和概念是基于正交几何体可连续变换的立体镶嵌。这些镶嵌的转化、结合和编辑创造了新的处理房屋构件的方式，如窗和门的开洞、建筑的内外区分、墙的细分等。在更小的尺度范围内也使用类似的几何镶嵌规则进行细分转化，以此生成洞口或纹样等小尺度元素，这些小尺度元素作为结果通过视觉和触觉贯穿于整个建筑。

In this project for housing in the city of Troy, New York, formal and conceptual development is based on processes of successively transforming three-dimensional tessellations of orthographic geometry. Transforming, combining, and editing these tessellations created novel ways of dealing with disciplinary issues such as window and door openings, interior vs. exterior, thickness of walls, etc. Subdivision and further transformation of the tessellations were used to produce pattern from the same geometrical system at a much smaller scale. Perforating and/or tattooing larger scale elements with smaller scale elements results in a range of visual and tactile affects throughout the building.

物化证明 /
Objectified Alibis

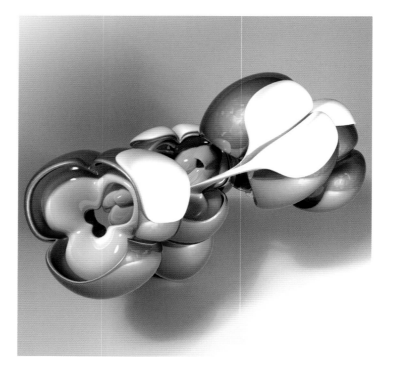

这是一个仪式性空间的构成设计。在最初,我们对巴洛克建筑进行全面分析,研究观察建筑物参数上的细微差别。通过对巴洛克建筑内部体量的分析,采取一个内外转换的方法,产生出许多不同的建筑场景。经过层压、重叠、分层、虚实转化后,内部围护结构和空隙空间被创造出来,复杂的多层次系统也被建立起来。从二维和三维去研究空间,并发掘这些空间与仪式队列的共同性,以此获得不可预期但却可加以控制的结果。

The studio undertook the task of designing a ceremonial space initiated by a comprehensive analysis of Baroque architecture. This entailed investigating and speculating on the parameters and nuances of the monument. Design experiments questioned a variety of architectural scenarios through an inside/out approach, beginning with the analysis of interior volumes found in Baroque buildings. With architectonic strategies of lamination, overlapping, layering, poché, mass and void, interior envelopes and interstitial spaces were created. Intricate multi-layered systems established spatial configurations such as inside/inside, inside/outside and outside/outside. These volumetric conditions were studied in 2D and 3D in order to find symbiosis with the processional program of the monument and to achieve an unpredictable but controlled outcome.

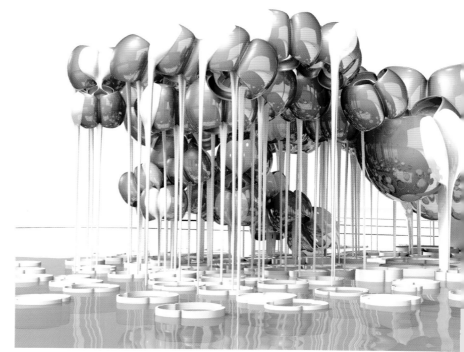

ERSIDE RENDERING // INTERIOR ENVELOPES

美国南加州建筑学院 /SCI-Arc

自命题 /
Self-Titled

将作为项目的数字化、作为工具的数字化以及数字化本身加以区别。这一行为的实现并未通过自身系统的物化，而是通过其明显的矛盾。因此，形式中动与静共存，连续与离散共存，与数字项目的创作趋近又背离。此外，既肯定厚度的存在，又否定数字化零厚度的表面。同时，比喻式的"彩色浸染"是对整体、形体边界以及表皮的肯定。最后，将三维表示为一个平面，或者说是将平面带入到三维。因此，该项目是一种成像，而不是创造。

This project looks to separate the conflations of the digital-as-project, digital-as-tool, and digital-as-end-in-itself. It proposes doing so not through the reification of its own system, but through its apparent contradictions. Thus form is seen as something both animate yet in repose, contiguous yet broken – as a nod to and departure from the originating digital projects. Moreover there is a tectonic division towards the affirmation of thickness, as a denial of the digital zero-thickness surface. Meanwhile there is a figurative 'color-dipping' as an affirmation of the whole, the outer boundary of the form, and the affirmation of surface. Finally, 3D is represented flatly, or rather the flat is brought into the 3D. The project is therefore seen as one of imaging and not production.

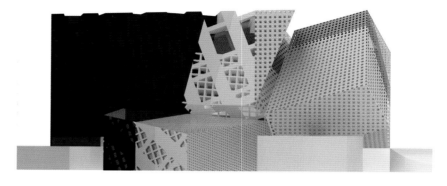

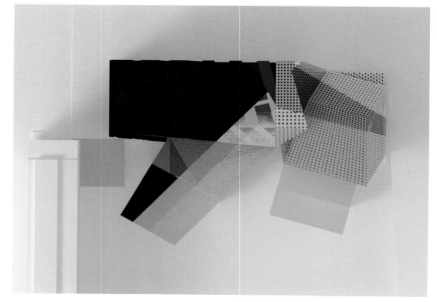

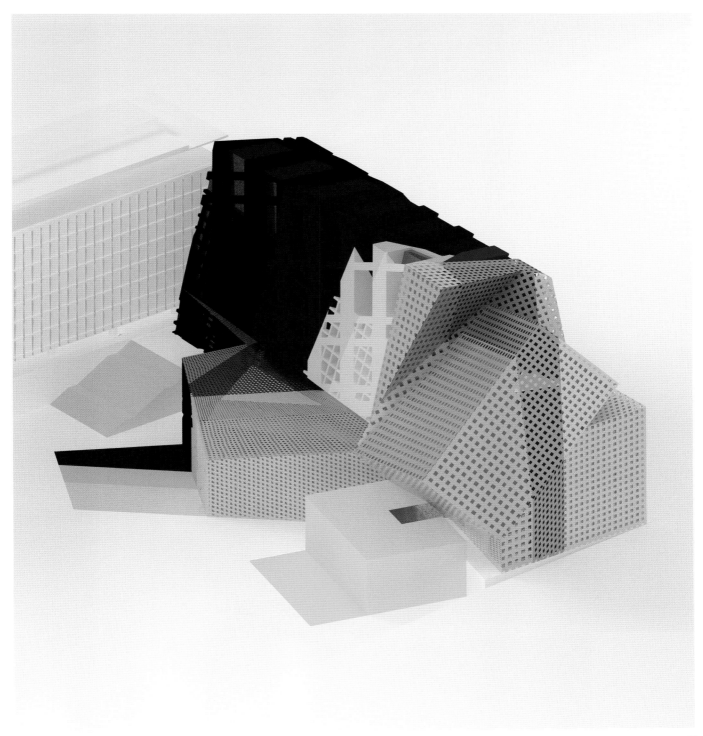

从表皮到花朵 /
Surfaces to Flowers

在一个数字化设计工具的使用已经趋同的时代，需要重塑以进行个性化表达。创造出一种新的融合，可利用冲孔操作形式和雕刻技术来实现内部和外部空间之间的互动。将动画电影技术作为设计工具，形成并提高空间的感知，同时在虚幻与现实中创造一种不安的模糊。这个项目包括一个艺术家的住宅以及维也纳会议厅的改造。它侧重于模拟令人不安的重物质，同时产生新的联系而不是矛盾元素。

We are in a moment where the use of digital tools for design has reached a level of homogeneous precision that needs to be corrupted and crafted to communicate individual expressions. There is an emergence of new coherences out of contradictory elements, using the technique of piercing for manipulation of form and carving as a way to generate interaction between the interior and exterior space. Animation and cinematic techniques are used as design tools, manipulating and enhancing the perception of the space, while creating an uncomfortable ambiguity between fiction and reality. This project consists of a house for an artist and transformation of the Secession Hall in Vienna. It focuses on the simulation of disturbing heavy matter, while generating new relationships out of contrasting elements.

画家般的数字化 /
Painterly Digital

计算机图像实现了如画般的特定效果。数字化创作的图像以相似的分层逻辑进行创作，并有着传统绘画的特质。这个设计通过利用 Processing 软件编写代码来处理数据。代码能模拟绘画的物质特性，并在数字环境中体现出来。单项的操作者编辑图形的像素序列来创造复杂的图像。

The impact of computer based imagery achieves a unique painterly effect. Digitally created images have the tectonic qualities of a traditional painting and are constructed within a similar layer-based logic. This project deals with the data bending process through a combination of written scripts in Processing. The scripts aim to adopt the physical qualities of paint and to perform them within a digital environment. Unary operators edit the pixel orders of an image resulting in complex graphics.

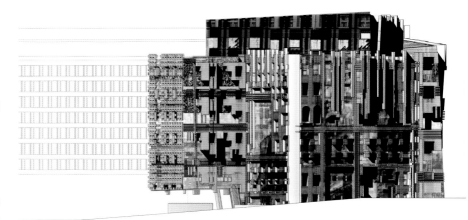

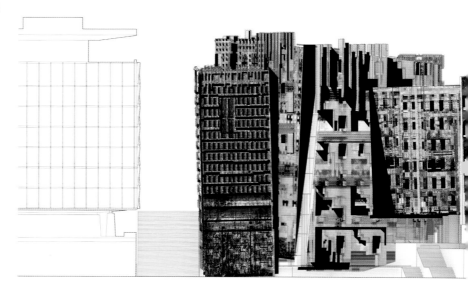

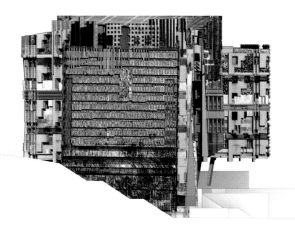
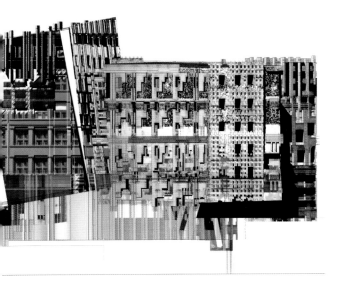

美国加州大学洛杉矶分校建筑系 /UCLA

瑞士军刀式机翼生产线 /
Swiss Army Knife Rooms

该项目旨在利用数字设计与建造技术为波音 777 飞机的机翼组装线寻找一种新的安装系统。类似瑞士军刀的结构，在一组系统中有四个房间，每个房间分别持有一架机翼，系统每发生一次动作，这四个房间位置就会按照次序依次转换到前一个房间。机翼在每一个房间进行不同的装配。持有机翼的房间留有开口，在白天为厂房提供采光；同时，这些房间在一个周期的动作中也会穿过出屋顶的天井，并形成一个可供游客参观的屋顶展厅。

This project looks at a new tooling system for the sub-assembly line of 777 Boeing airplane's wing by the use of advanced digital techniques in the design and fabrication. Similar to a Swiss army knife, there are 4 rooms holding the wing that replace each other through each motion. The wing is assembled in each of the positions. The rooms holding the wings include openings that allow daylight into the factory. The moving room which holds the wing also comes out of the ceiling during its cycle, moving through the skylight to provide a rooftop showroom for visitors.

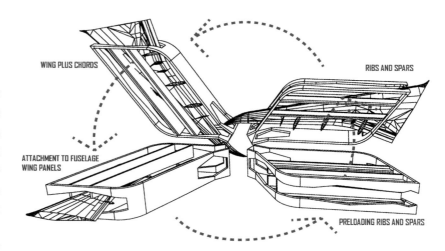

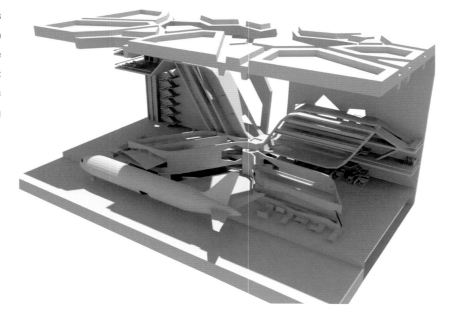

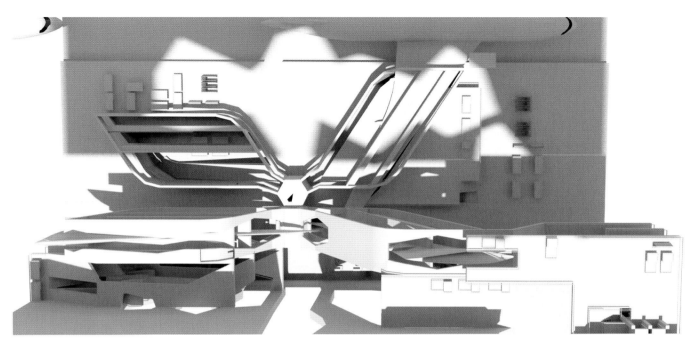

多孔块体 /
Porous Mass

该项目研究多孔物质，从实体与空腔之间的界限着手，尝试去界定一种丰富的、界于虚与实、深与浅、内与外的第三状态，打破建筑现有的二元化的趋势。最终，建筑将是一个单独的整体，而非罗朵尔福·马查多所说的其特点仅反存于表皮厚度的"表皮的整体"，该项目由一个经过溶解一定时间形成的连续多孔的物体构成空间。它的剖面指向一种增厚的空间形式，形成一种既非外亦非内、既非表面亦非实体的空间状态。

This project is an exploration of porous mass, operating between the limits of solid and hollow, to locate a fecund third condition which disrupts architecture's tendency towards the binary, a condition between solid and void; surface and poche; interior and exterior. Ultimately, architecture can only pretend to be monolithic. As opposed to Rodolfo Machado's classification of an 'Epidermal Monolith,' whose monolithic character is registered within the thickness of the envelope, this project consists of a continuous porous mass with dissolves in certain moments to form space. Its section points toward a thickened form, and explores an architectural condition born of neither surface nor solidity – neither exterior nor interior.

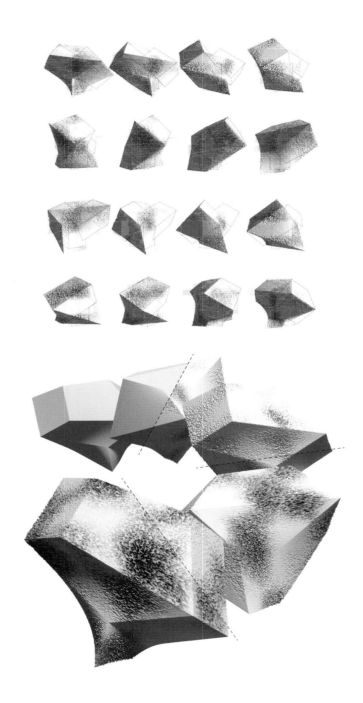

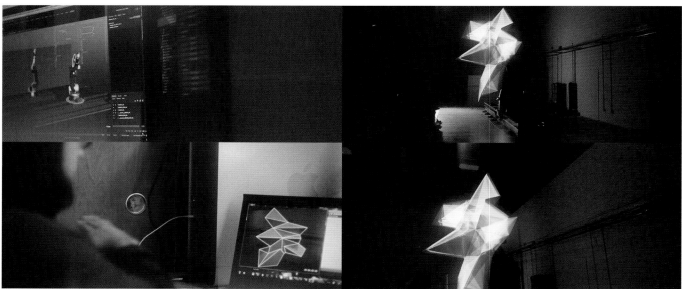

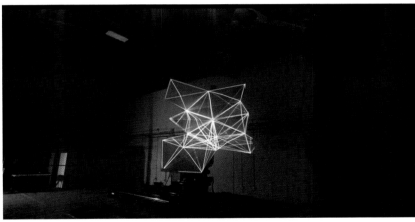

以太项目 /
The Aether Project

以太项目的重点是通过一个实时的跃动控制系统，将视像投影到一个可以按照编排好的机器人动作做出变形的几何表面，实现一种拟真体验。该项目旨在测试人与机器人、机器人与机器人之间的互动关系，以及技术与人的认知能力之间不断递归循环的关系。该项目在上述技术与理论中找到一个结合点，即通过传感技术实现空间环境对于其居住者的感知。通过采集人的信息（手势）作为输入信息，并通过编译的机器人的动作将其转化为实时的形态变化。

The Aether Project focuses on providing an immersive experience through real-time Leap Motion controlled system that synchronously aligns projection mapped visuals on a transforming surface geometry, both choreographed though robot movements. The project is designed to test the interaction between humans and robots, robot and robot, and the resulting recursive relationship between technology and human perception . Mediating between aforementioned technological and theoretical paradigms, the system culminates into a spatial environment that is conscious of its inhabitant(s) through sensing technology. It achieves this by taking human inputs (hand gestures) and translates them into real time formal deformation through choreographed robot movement.

美国密歇根大学 /UMichigan

倒模 /
AL-10XX

本项目探究依赖于受物理限制的铝工件设计生产过程。步骤包括：制作木模，沙模翻模，机械抛光，机械铣模，浇注熔融铝水及后期制作。通过在原始流程中加入使用7轴机械臂和先进的软件技术，我们实现了数字模拟的跨越衔接。我们本次的研究旨在探索未来的、创造性的制造流程。我们尝试去调和审美意义上的完美的与不完美的、设计的与突现的，而在这一过程中我们为建筑师和设计师找到了一种新的可能的角色定位，即"数字工匠"和"自然过程的陪护人"。

This project explores an aluminum artifact design process dependent on tangible physical restraints. The steps involved included wood framing, sandcast molds, robotic burnishing, robotic scratching/milling, pouring molten aluminum, and post-production. Employing the 7-axis robot at the intersection of primordial elements and cutting-edge software, the project navigated the gap between digital and analog. The research is intended as a precursor to future creative manufacturing processes, reconciling such paradoxes as the aesthetically perfect vs. imperfect, and the designed vs. emergent. In so doing it discovered truly unique results, suggesting a new possible role for architects/designers as "digital craftsmen" and "chaperones of natural processes".

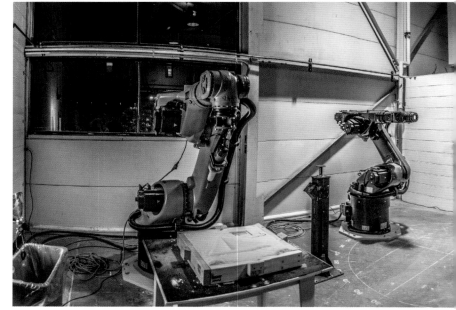

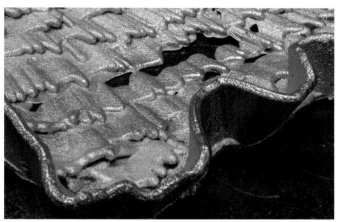

界定 /
Domain

本项目始于对于历史案例、当代空间概念理论和原则的一个研究课题组,后期发展为真实的设计与建造工作营。我们仔细研究了双曲形态在结构、声学、美学和文化方面的表现。随着设计的改进,实体模型尺度增加,我们尝试在不使用模具的情况下建造一个双曲壳面。沿着历史轨迹,尤其是巴洛克建筑中的曲线和几何形态,项目开发了一个设计程序的脚本,并与材料、结构和建造三个方面对接,以完成该课题。

This project was part of a research studio based on the analysis of historic examples and contemporary theories and principles of spatial concepts to develop into a design and manufacturing studio. Double curvature was scrutinized for its performative qualities in the fields of structure, acoustics, aesthetics, and culture. Accompanying the design evaluation physical models were fabricated increasing in scale, attempting to create a double curved shell without molds. Continuing a historic trajectory, specifically the application of curvature and geometry in Baroque architecture, the project developed design protocols that interfaced material, structural and manufacturing aspects.

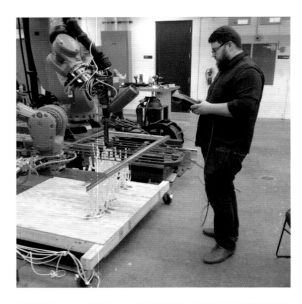

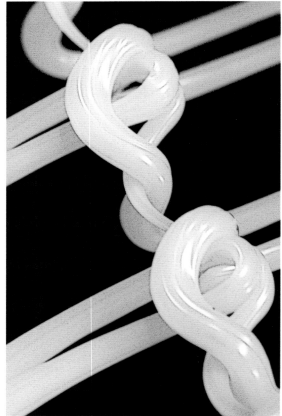

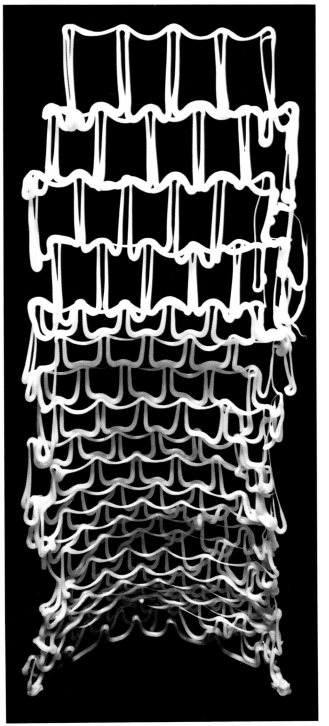

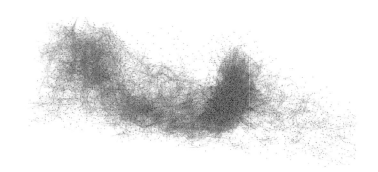

代理智能模拟 /
Agent Intelligence

代理计算是一种模拟代理系统运动方式的计算机数学模型。在该项目中,我们将这种运动通过一组基于材料特性和制造约束的程序规则,转译成钢杆的几何形态。在这组规则的框架内,不同规则的组合将产生不同的变化。在设计加工过程中,我们采用了一种集成了生成式设计、材料特性及制造的反馈网络。我们使用机器弯杆和数控加工技术接受脚本,并将数据返回到设计程序,从而提升设计加工的品质。

Agent-based computing is one computational model for simulating the movements of agents. In this project these movements were interpreted into the geometries of steel rods via a set of programming rules translated from the material properties and fabrication constraints. Within the frame of these rules, variations could be generated through different compositions of the rules. During the design-fabrication process, the project adopted an integrated feedback network of formational design, material properties and fabrication concerns. It used the robotic rod bending and CNC machining techniques to receive scripts and return feedback to the design process, so as to improve the design-fabrication process.

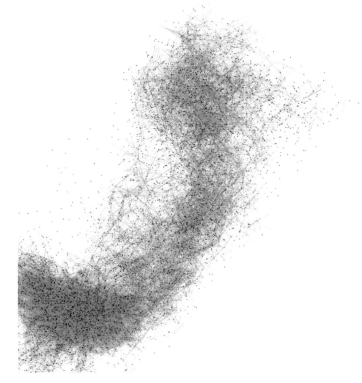

美国宾夕法尼亚大学建筑系 /UPenn

黑洞 /
Black Holes

原始独特的洞通过减缺或者表面拉伸的方式由厚实的形体转变为透明或者类似的形体。这些推进、推出或者不动的图形的相互作用,创造出神奇的图形,这些图形尽管看起来好像有很多"通道",然而却在藐视穿越。就像黑洞一样,这些内陷的图形可能并没有一般意义上的入口,反而是一种持续不断的吸引,并不存在真的通道。亦真亦假的阴影、反射和光晕成为真实的物理特征,这些特征以表面作用的形式存在,比如材料或者表面光泽度的改变。它们创造了扁平化、深化、模糊化等作用,这些作用增强了建筑物的神秘性和模糊多样性。

Strange, primitive holes are made from either subtracting or pushing chunky figures into crystalline containers or into one another. The interplay between figures which push out, push in, or remain hidden create mysterious formations which defy access, although they may appear to have multiple 'doorways'. Like black holes, these involutions may not constitute literal points of entry but rather moments of allure and lack of access. Fake and real shadows, reflections, and halos are reified into physical features. These features exist as surface effects, such as changes of material or sheen. They create flattening, depth, or obscuring effects that heighten the mystery and irresolution of the building object.

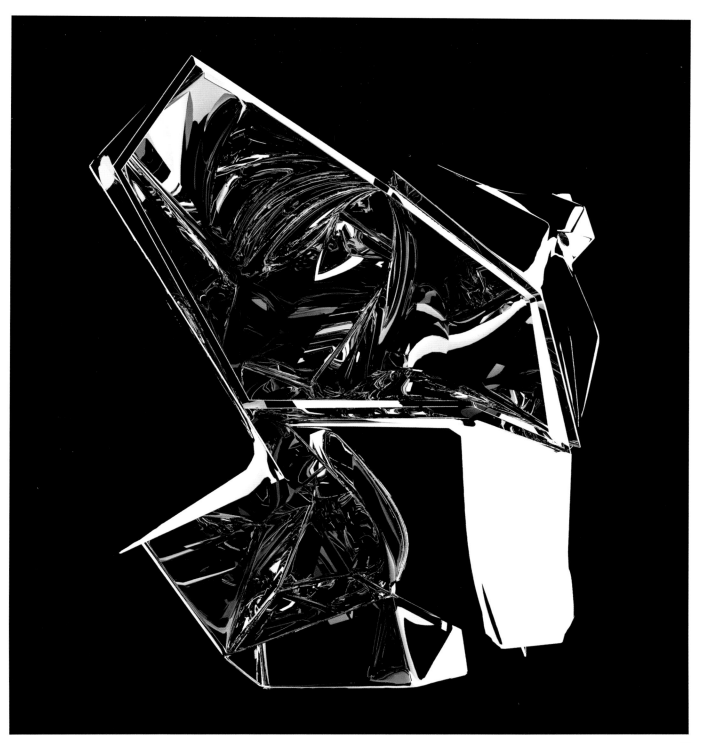

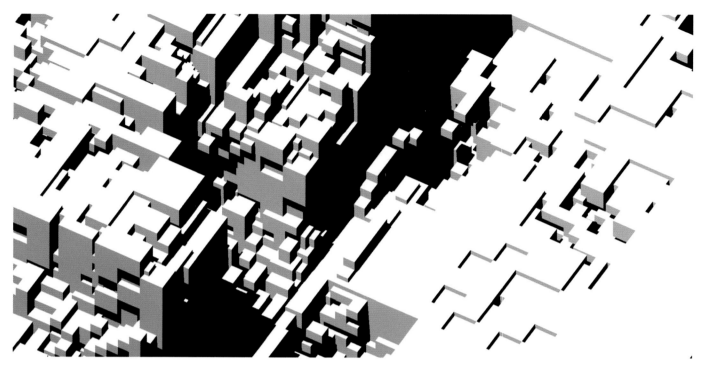
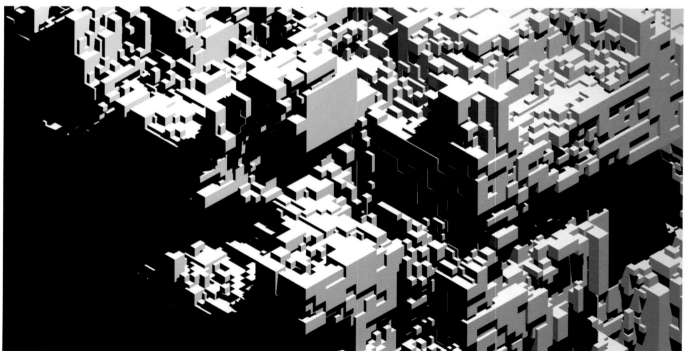

幻想图书馆 /
The Unimaginable Library

在豪尔赫·博尔赫斯短片故事《通天塔图书馆》中，他给我们描述了一个神秘的世界，这个世界由一个巨大的图书馆构成，里面收藏了所有 410 页长的图书。每一本书的每一页都有 40 行，每一行都有 80 个字，且包含了所有 25 个可能的字母符号。他试着用计算机生成这个看似不可能的情况。该项目背后的基本假设，首先，计数是思考的基础；其次，计算是离散的；最后，现实是以离散的形式存在。

In his famous short story about the Library of Babel Jorge Luis Borges gives us an enigmatic description of a universe comprising a vast library consisting of all possible books that are 410 pages long. In this proposal each page of each of these book is assigned 40 lines each consisting of 80 positions, drawing from 25 possible alphabetic characters. An attempt is then made to use computation to generate possible outcomes to this seemingly impossible scenario. The postulates behind the project are firstly that counting is fundamental to thought, secondly that counting is discrete, and thirdly that reality is in a form of discreteness.

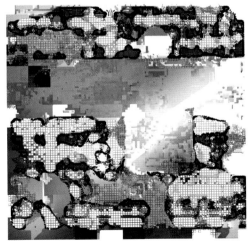

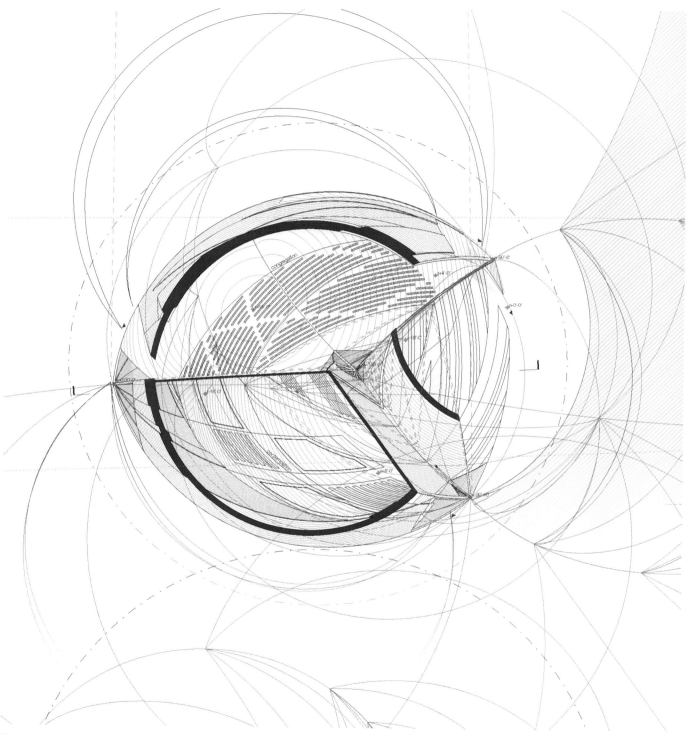

无穷小 /
INFINITEsimal

第四个空间维度可以被描述为三维世界在垂直方向的延伸。因此，我们的现实世界是多维时间空间的一个极小片段。我们使用不断细分的方式在有限的空间内创造虚拟的无穷。这些相似的几何形不可避免地压缩成为极小的点。为了感知到四维空间，可以用等高线划分或者其他划分方式表达。立体画法中的"极远处的点"创造了一个消失的地平线，所有的几何形都将在那里终结。

The fourth spatial dimension can be described as an extrusion of our 3D world in a direction perpendicular to the perceivable three. Thus, our reality is merely an infinitesimal slice of the compounded folds in space-time. Recursive subdivision may be used as a method of creating a virtual infinitude within a finite geometry. These self-similar geometries inevitably converge to points of infinitesimal size. In order to perceive a four-dimensional geometry, one may utilize contours or other projection methods. The "point at infinity," used in stereographic projection creates a vanishing "horizon" as the geometry approaches the point of projection.

美国南加州大学建筑学院 / USC

Dailemota/
Dailemota

多联骨牌2.0探索游戏技术在设计中的使用，专注于算法和人类玩家之间的共生的组合模式开发。这个项目开始于对一个完美镶嵌的几何形态和一个立体单元的探索，但是通过打破一个多面填充体的空间对称性，并允许在尺度上循环迭代，Dailemota项目能够打破对称单元序列的单调性并通过人类设计师进行不同的构造。多联骨牌代理是由"从游戏到制造"研究建立的框架，这项研究的目的在于链接游戏技术和3D打印技术，允许非专家的玩家进行群众外包设计。

Polymino 2.0 explores the use of gaming technology for design, focusing on ideas of combinatorics and patterns develop by a symbiosis between algorithms and a human player. The project begins with an exploration of perfect packing geometries and a voxel, but by breaking the symmetry of a space filling polyhedral and allowing for recursive transitions in scale, the Dailimota project is able to break the monotony of a symmetric voxel array and differentiate tectonics in the interest of the human designer. The Polyomino agenda is framed by the 'Gaming to Making' research that aims to connect gaming technology with 3D printing, allowing the possibility of crowdsourcing design in the hands of non-expert players.

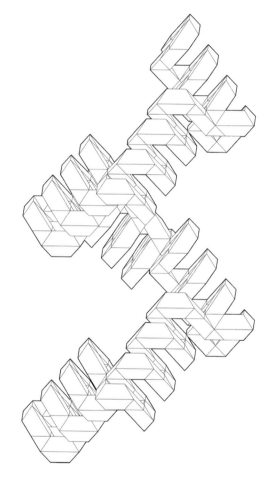

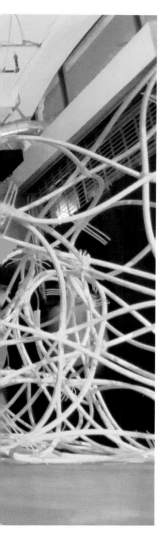

爱丽丝的仙境 /
Alice in Wonderland

爱丽丝的仙境是由本科生设计的一组在洛杉矶市中心的互动装置。这个项目试图捕捉路易斯·卡洛儿书中"超俗"、"神秘"的角色，一个关于爱丽丝的迷人的故事。爱丽丝是一个掉进了兔子洞、进入了魔幻王国的小女孩，一系列的场景由此产生。这个装置组合了一系列独特的干预机制，包含互动的花朵、像脊柱的尾巴和充气硅结构，聚集在像丛林一样的一堆管子周围。这些独立的装置通过由 Arduino 控制板控制的运动传感器来接受刺激，之后通过可视化编程语言、MaxMSP 或者 Processing 软件来实现。

Alice in Wonderland was a group interactive installation for Downtown Los Angeles assembled by undergraduate students. The project sought to capture the 'other-worldly', uncanny character of Lewis Carroll's enchanting story about Alice, a little girl who falls down a rabbit hole and enters a magical kingdom, on which a series of movies have been based. The installation comprised a series of individual interventions consisting of interactive flowers, spine-like tails, and inflatable silicon structures, clustered around a jungle-like mass of tubing. The individual installations were actuated through motion sensors controlled by Arduino control boards, and were programmed using either the visual programming language, MaxMSP or Processing software.

洛杉矶当代艺术博物馆 /
LACMA++

洛杉矶当代艺术博物馆设计项目调查洛杉矶现有的博物馆，探讨了未来的博物馆。有目的的调查包含了程序和构造技术。在技术层面，这个项目通过一系列的参数化的、生成的和3D打印、CNC加工等技术，用明确的立体模型技术探索了数字原型。这个项目呈现出一个自定的运用MAYA和Rhino-Grasshopper的工作流程，是一个具有复杂曲率和离散形的高精度几何形的范例。

LACMA ++ is design project that investigates the Museum of and for the future for Los Angeles. Purposefully the line of inquiry is of programmatic and tectonic hybridity. Technologically the project explores digital prototypes through a series of parametric, generative, and explicit modelling techniques along with 3D printing and CNC fabrication. The project presented is developed through a customized workflow using MAYA and Rhino-Grasshopper, and is an exemplar of geometric precision with complex curvatures and discretization of forms.

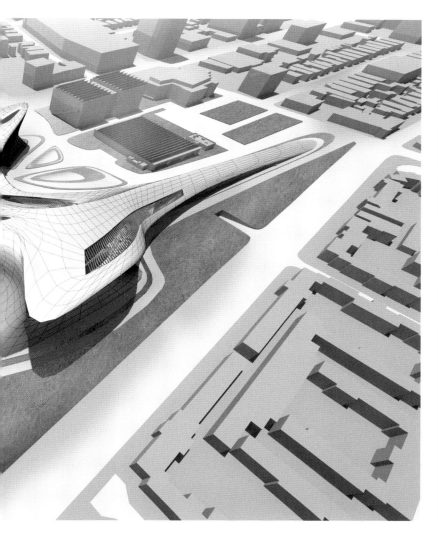
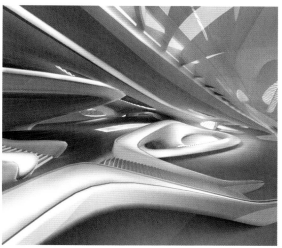

美国耶鲁大学建筑学院 /Yale

图样主义 /
Patternism

现在参数化图样表皮、空间网格、幕墙等比比皆是。除了理性的饰面设计与合理的操作,参数化大多数情况下是一种图样执行技术,而不是空间制造工具。本方案尝试挑战这种价值观,从二维图样出发,进而去生成具有一定体积的一系列空间。从罗伯特·文丘里关于复杂性和矛盾性的想法出发,方案生成了大量具有多层内部体量的物理模型,这些模型具有强烈的反差,并且嵌套了多种层级与对称构成。

Parametrically patterned surfaces, space frames and curtain walls now abound. Despite a veneer of rationalized design and verifiable performance, this is still a set of techniques valued more for their graphic performance than their space-making capability. This project attempts to invert that set of values, starting with graphic, 2-dimensional pattern but expanding it to produce volumetric matrices of space. Working from Robert Venturi's ideas about complexity and contradiction, which he read through 2D drawing, the project developed large physical models with multiple interior volumes defined by sharp contrasts, nested hierarchies and local symmetries.

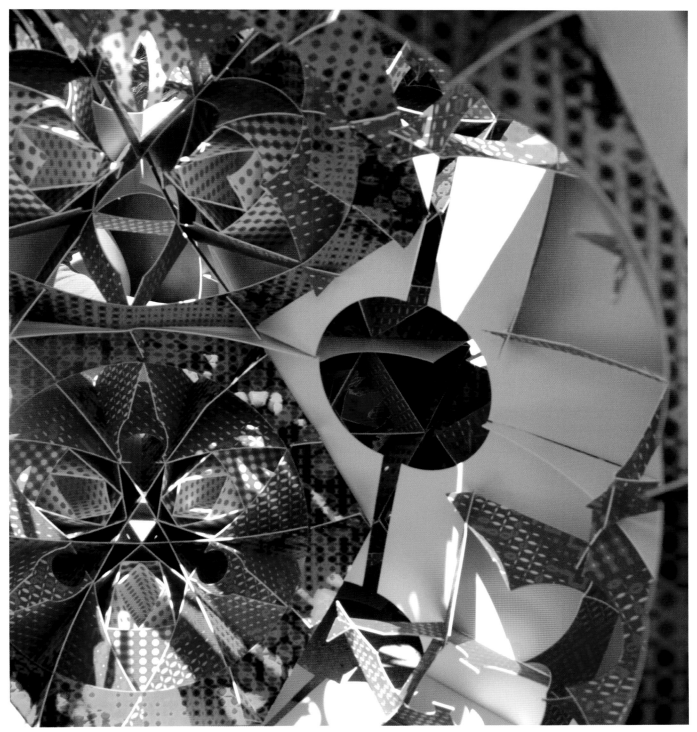

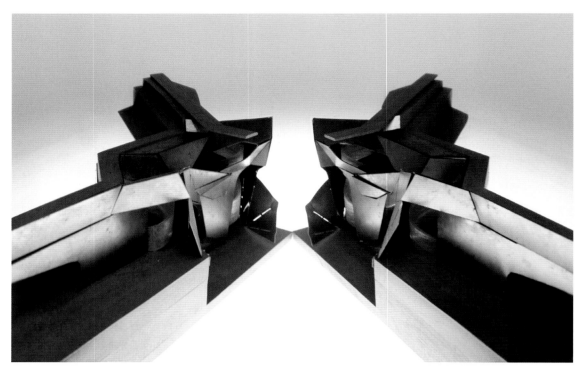
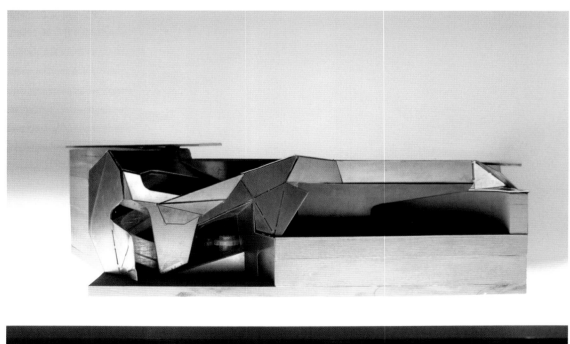

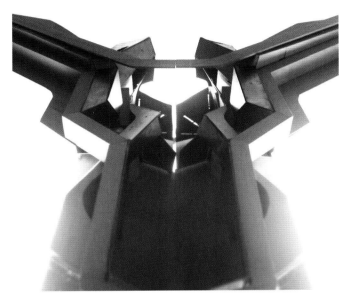

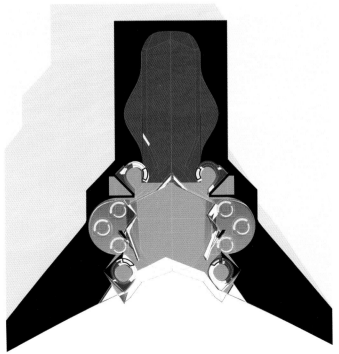

硬壳 /
Carapace

本方案利用水闸去显露原本应该高度密封的功能：水渠、溢洪道、动力室、涡轮大厅，进而强调了大坝的城市特性，通过水闸的内立面去暴露它们，就像蓬皮杜中心一样。将"地下"转换为"水下"，大坝将这种体验夸张化，将上游与下游之间以及其内部的各种元素显露出来。用动画软件来模拟上述过程，利用数字化语言去模拟折叠表面与光滑实体；然后利用数控系统成型机和模拟技术将数字模型转化为实体模型。

This project addresses the civic character of the dam by using the lock to reveal what would otherwise be highly internalized functions – the penstocks, the spillways, the powerhouse and the turbine hall – by exposing them throughout the internal façade of the lock, similar to the Pompidou Center. Translating underground into underwater, the dam dramatizes this experience by submerging and revealing its elements within the space between the upstream and downstream of the river. Animation software was used to simulate this experience within a digitally generated language of folding surfaces and smooth solids. A combination of CNC prototyping and analog techniques then moved from digital model into physical form.

凌乱的几何 /
Disheveled Geometries

如果历史上的乡土建筑被视作以不够精致的成型材料建造，以致肌理粗糙，那么 21 世纪的情况则与之完全相反。在本方案中，学生们学习历史中建筑乡土化的方法，并且将这些调研作为基础，通过 Kitbashing 技术，设计和制造出一个独立的，乡土化的华丽 3D 打印体。这个方案的实现是综合运用了 Thingiverse, Autodesk Maya, Fusion 360, and Mudbox 等软件进行工作，由 1220mm x 900mm x 610mm 的实体进行打印，然后数控铣削，材质是大理石碎片。

If, historically, architectural rustication was seen as a less refined manner of shaping material that subsequently retained a rough texture, then the 21st century condition would be the exact reverse. In this project, students studied methods of rustication throughout history and used this research as a foundation to design and produce, through techniques of kitbashing, a single, heavily figured and rusticated 3D printed volume. This was completed through a workflow of using Thingiverse, Autodesk Maya, Fusion 360, and Mudbox and culminated in a solid 1220 x 900 x 610mm 3d printed object and a CNC-milled, marble fragment.

英国建筑联盟建筑学院 /AA

灌入 /
Infusion

本方案设计出一个灵活、自动的机器制造系统，它能够利用机械手臂，采用可降解材料进行精准的重复性动作工作。这个系统允许连续的成型格子结构能够溶解，材料能再利用或者进行生物降解。这种可重复利用的活性机制能够运用在现代建筑的各个领域，特别是可以利用温度进行现场人工组装。整个系统没有节点，并且采用单一的材料，同时能够自承重，因此其也不需要模板材料。这个制造系统改变了现有的、死板的、工厂预制的制造模式，利用可降解材料和现场机械制造，实现了定制化建筑。

This project proposes a flexible, automated robotic fabrication system, which exploits phase-changing materials and utilises the intelligence of a robotic arm through accurate repetitive actions. The system allows the fabrication of a continuous lattice structure that can be melted and reused or left to biodegrade. The dynamism and reusability of the system results in a variety of temporary architectural applications, where the manufactured lattice is manually assembled on site using heat. This project proposes a reusable, biodegradable and temporary architecture. The system is a joint-less, mono-material, self-supporting system that does not require formwork. The fabrication system is enclosed in a mobile unit, shifting the existing paradigm of the factory away from rigid pre-fabricated components and towards custom architecture utilising phase-changing materials with on-site robotic fabrication.

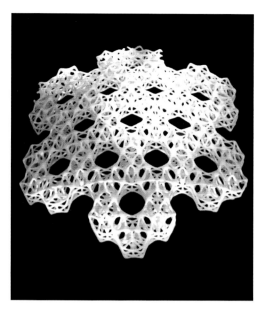

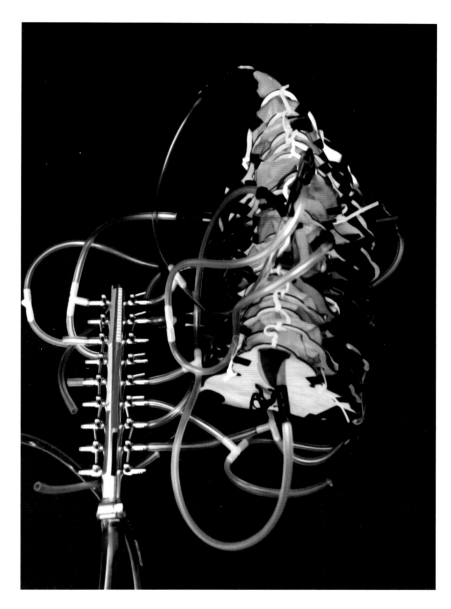

OwO/
OwO

OwO 为城市环境提出了一个新的模型，一个基于简单规则和复杂关系以及通过两者相互关系的行为系统，这个新模型能够自我组装，自我成型，并且随着时间的演进进行自我更新和再造。这个提议模糊了人工制造与自然的明确界限，并且使得基于简单的群体行为就能自然生成空间组织有了可能。本方案将自身整合为一个公共空间模型，它是一个灵活并且能自我组织的，在不同情况下可执行不同尺度要求的系统。由此，曾经固定且永久的公共环境转变为了可适应众人分享需求的公共空间模型。

OwO is proposing a new model for the urban environment; a behavioural system based on simple rules and the complex relations and interactions between them in order to achieve a model of self-assembly, self- structuring, adaptation and reconfiguration in time. The proposal blurs the distinction between the artificial and the natural and opens up the possibilities of emergent space organisations that are based on singular and group behaviours. The project integrates itself into a model for public space as a flexible and self-organised system that operates on different scales and conditions. It proposes a shift from a fixed and permanent public environment, towards an adaptable and participatory model for public space.

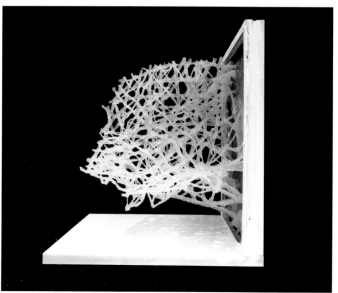
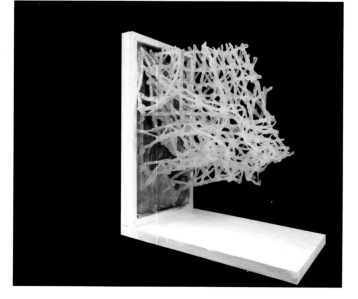

空中打印填充 /
Aerial Printed Infill

该研究依据机械系统的行为模式以及使用复合材料，提出了一种由一群无人机参与的非标准化建造过程。系统驾驭无人机进行按预先设计方案建造的同时，也能根据环境的反馈情况，灵活地改变建造策略。研究提出了一种基于时间的，可自适应的，实时更新的系统机制，能够让材料组织、形式、结构、尺度、构成根据所处环境、天气、结构与材料的呈现以及其他不可预估的场地限制因素的回馈，而做出适应。

The research proposes a non-standard construction process created by a swarm of UAVs. This process merges design and production within a singular process, based on the behaviours of robotic systems and composite materials. In contrast to conventional construction techniques, the system harnesses drone technologies in order to propose a construction process that can be pre-designed yet allow for flexibility in order to be implementable within a variety of environments through real-time structural feedback. The research argues for a time-based system adaptive process that is flexible in material organisation, form, structure, scale and configuration that is capable of responding to fluctuations in local environments and weather, structural and material behaviour and other unpredictable site-constraints.

奥地利维也纳工艺美术学院 /Angewandte

自构领域 /
Sematectonic Fields

这个方案是基于这样一个前提：建筑师不再是空间和物质的组织者，而是在复杂"工程生态"环境下的一个多层组件的系统设计师。它着力于生成一个自下而上的城市形成系统，这个系统可以自我学习、构建、适应、进化，利用自我聚集的结构作为生成过程，去动态控制当地的信息，进而去影响更大范围内城市生活的过程。它探索了行为设计方法的潜力，以算法为媒介去生成新的高密度、多功能的空间原型。

The project is based on the premise that the architect is no longer an organizer of space and matter, but rather a designer of systems with multi-layered components in complex "engineered ecologies". It targets the production of urban formations as a bottom up system that has the capacity to learn, self-structure, adapt, and co-evolve, using self-organized swarm structures as a generative process to control local dynamic information to influence larger urban life-processes. It explores the potential of behavioral design methodologies through the medium of computation to generate new typologies of high density poly-functional spaces.

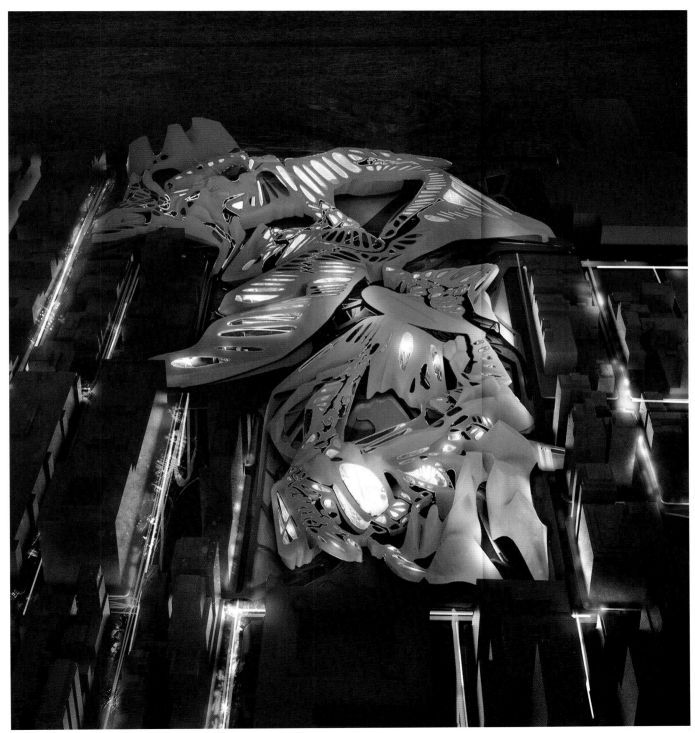

展开的巨石 /
Unfolding Monolith

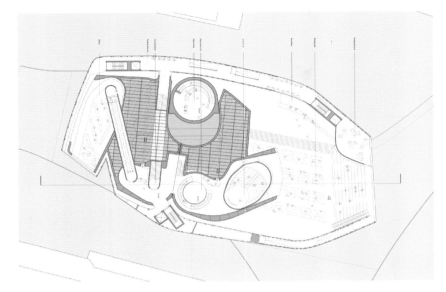

这是一个复合功能,运动学连接,具有可移动房间的项目,它能改变自己内部的构造以及空间的原型,同时也能保持建筑具有一个固定的外壳。这种空间的转变是通过其内部"巨石"的展开和收拢得以实现的,由此改变内部空间之间的关系和连接方式,这些活动都是在建筑具有一个固定单一的形体的前提下发生的。方案的关注点在于两个参与变化的空间关系:内部的房间,它们只基于相邻的房间进行线性的变动;大厅,由容纳这些房间的建筑体界定。

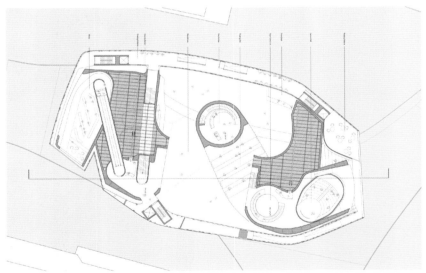

This project is about multiple, kinematically connected, moving rooms that can transform their interior configuration and typology while maintaining a single figure. This transformation occurs through the folding and unfolding of a monolith, changing the relationship and connectivity of spaces within while transforming the posture of a single body. The focus lies in the relationship of the two transforming spaces: the rooms, which transform linearly only in regard to their adjacent rooms, and the hall which is defined by the transforming figure (the body containing the rooms) and the casing (a surface in a loose-fit relationship with the moving body).

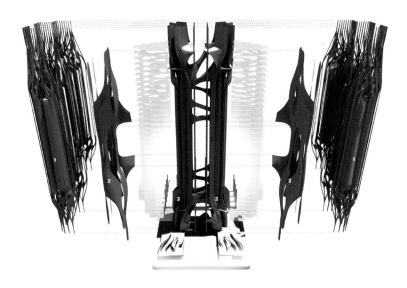

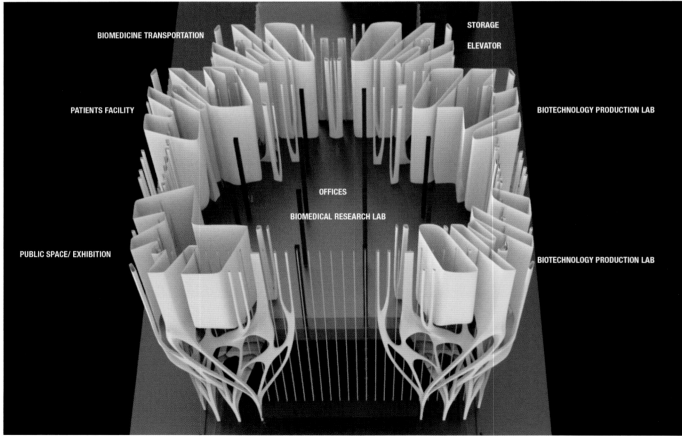

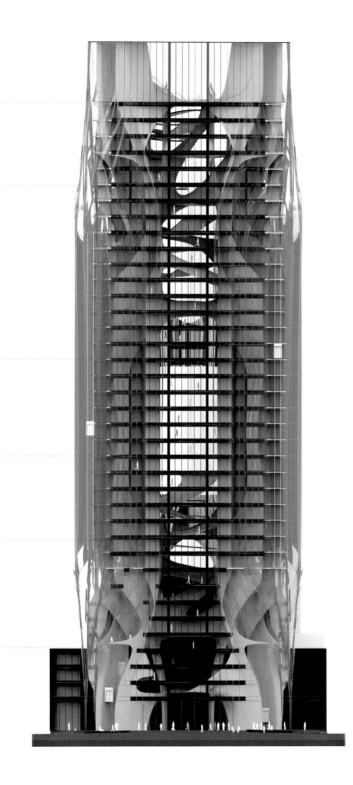

H+/
H+

H+ 是纽约一家生物工程公司的总部，用于进行产品研究和展示。方案受人体启发，再经过生物以及机器的优化，大楼的核心是"累加"。利用I型梁的特性，构成了附加的空间并作为产品实验室、公共设施以及垂直交通；同时，核心部分的楼板则从周围的隔墙中解放出来。新的结构形态位于建筑的周边及角落，它们拓展了自动垂直实验的技术视角，展现了公司的业务和研究进展。

H+ is a new headquarters for a biotechnology company in New York, as a space for a research-production and exhibition area showing the latest developments within the field. Taking inspiration from the advancement of the human body through biological and robotic modifications, Mies van der Rohe's and Philip Johnson's Seagram Building is subjected to augmentation. The characteristic I-beam structure gets a volumetric addition which serves as production labs, public facilities and vertical circulation while the center of the slab gets liberated from the partition walls. The new structural morphology as shown in the six prosthetic-production towers situated on the corners of the existing building, explores the technological speculation of an automated vertical lab, revealing the company's activity and research developments.

英国伦敦大学巴特利建筑学院 / Bartlett

沙印 / SanDPrint

该项目尝试开发一项新的制作技术，用砂模来创造出各种独特的形体。选择PVC管是为了控制内部砂子的形状。在初步的材料勘察和制造测试之后，使用到了计算机技术。在数字开采的过程中使用了仿真加工技术，通过对从小规模组件到家具大小的规模进行研究，最终将设计扩展到建筑规模；同时，为设计提供了自然和逻辑的模式与结构。

This project attempts to develop a new crafting technique that uses sand as a mould to create a variety of unique forms. PVC tubes were selected in order to control the shape of sand inside. After initial material exploration and fabrication testing, computational technology was used. Processing simulation techniques (flocking, stigmergy etc.) were used in the digital exploration which provides both natural and logical pattern and structures by presenting products from the scale of small components to furniture, and eventually attempting to scale up the design to an architectural scale.

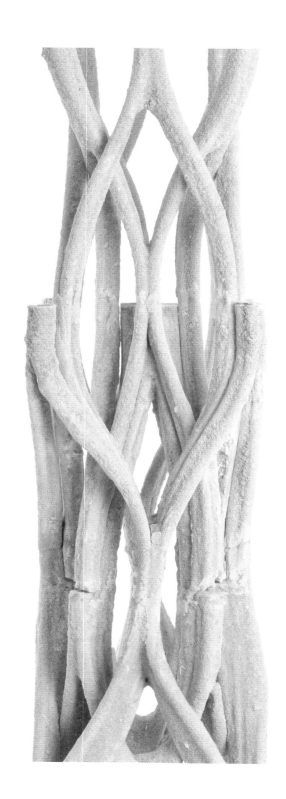

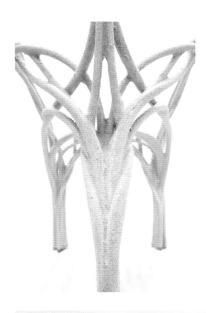

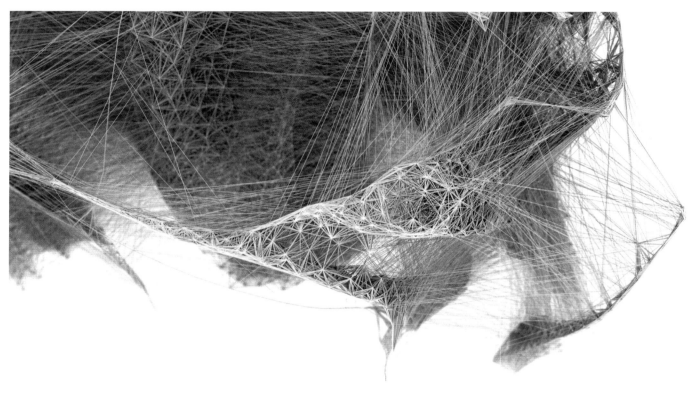

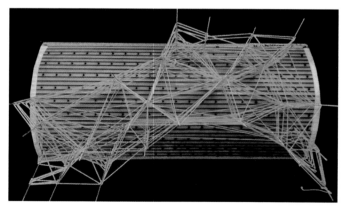

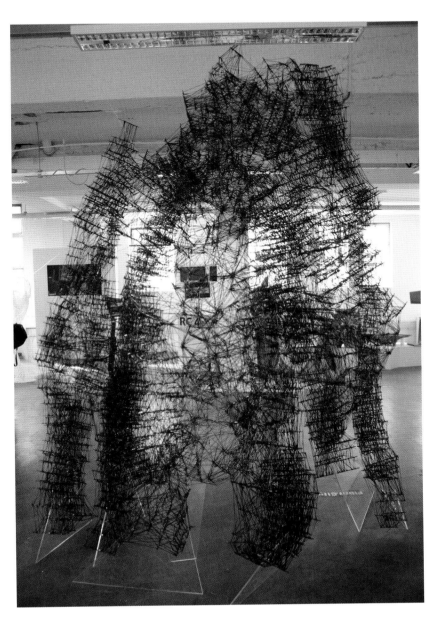

纤维城市 /
Fibro City

纤维城市是一个机器人编织碳纤维结构设计的研究项目。多代理系统应用于桥梁设计，以及材料和制造受约束条件下的设计。物理仿真用于获取高分辨率的结构优化结果以及复杂交错的结构网络。将信息丰富的算法层嵌入拥有设计和材料数据的调解处理层（从材料属性、机器人编织设计到结构优化）。编织代理系统将这些信息纳入其行为模式，所得到的结构具有很强的适应性、多样性并且轻质坚固。

Fibro City is a design research project on robotically woven carbon-fibre structures. Multi-agent systems are applied to bridge design intent and constraints of materials and fabrication. Physics simulation is used to derive high resolution structural optimisation and resultant intricate and interlaced structural networks. Information rich algorithmic layers are scripted into the process mediating layers of design and materialisation data (from material properties, constraints of robotic weaving choreography and structural optimisation). Weaving agents incorporate this information into their behavioural patterns. The resultant structures are highly adaptive, heterogeneous, lightweight and strong.

暗纹 /
Filamentrics

该项目研究将计算设计方法用于工业机器人的大型3D打印,将逻辑的、结构的、材料的限制作为设计机会,创造出具有高密度信息的非具象建筑空间。该项目旨在克服空间框架的限制,通过将机器人的塑料挤压引入设计和制造工艺,来避免同质性。可以在短时间内得到大量的具有高分辨率的打印结构和复杂的层次结构。定制的设计应用程序控制塑料压缩,将结构数据和一系列设计参数相关联,从而能够不断地为机器人制造刀具轨迹。

This project researches computational design methodologies for large-scale 3D printing with industrial robots, taking logistical, structural and material constraints as design opportunities to generate non-representational architectural spaces with a high density of information. The project aims to overcome the formal limitations of space frames, avoiding homogeneity by introducing robotic plastic extrusion into the design and fabrication process. Printed structures with a high degree of fine resolution and complex hierarchy can be achieved on a large scale within a relatively short period of time. The custom-made design application controls plastic extrusion, relating structural data to a variety of design parameters, allowing the continuous generation of tool paths for robotic fabrication.

丹麦皇家美术研究院信息技术与建筑中心 /CITA

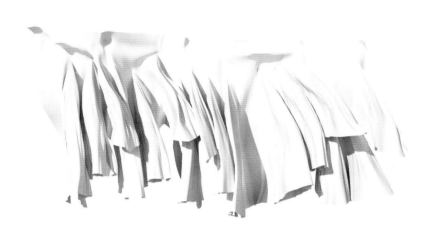

柔性建筑表面 /
Supple Architectural Surfaces

坐落在丹麦暴露的日德兰半岛西海岸边,这个项目探索一种建筑,可以因周边的环境而变得富有活力。使用易弯曲的鱼竿和纺织薄膜,并发展成为游客中心。通过这种建筑语言,在这里表达出室内和室外的关系有多层、扩展的界限。首先研究数字纺织模拟,然后进行设计,再指定和调整纺织膜的性能。纺织薄膜将成为虚拟程序和现实环境需求之间的界面。

Sited on the exposed West Coast of Jutland, Denmark, this project speculates upon an architecture that can be physically animated by its environment. Flexible rods and textile membranes are used to develop an architectural language for a visitor centre in which the relationship between interior and exterior is considered as a deeply layered and extended threshold. Digital textile simulations are informed through empirical studies and then employed to design, specify and calibrate a textile performance that acts as an interface between the demands of programme and environment.

特异性生长 /
The Growth of Specificity

这个项目发展了由计算驱动的演进设计策略,用来处理一个明确的规划和建筑类型(种子),它们会对欧洲的三个选定地点多样的气候、地理条件和标量需求做出反应(成长)。这是一个定位为具有生产酒背景的项目,并且需要在葡萄成熟期间为巡回劳动的员工提供临时住宿条件。这种临时居住的类型被扩展为一种特制的启动式建筑,并提供良好的室内控制环境和具有局部差异的空间,来适应一系列不同的活动需求。建筑方面的考虑涉及临时居住结构的设计和临时居住结构与葡萄园永久结构之间相对独立的关系。

This project develops computationally driven evolutionary design strategies for dealing with a specific programme and architectural typology (the seed) that can respond (grow) to varying climatic/geographic conditions and scalar demands across three chosen sites in Europe. The project locates itself within a programmatic context of wine production and the need to provide temporary accommodation for itinerate workforce during the grape harvesting period. The typology of the tent is expanded into a bespoke pneumatic construction that provides tuned environmental control and locally differentiated spaces to support a range of activities. The architectural consideration extends beyond the design of the tent structure and into the dependent relations that are established between the temporary and permanent structures of the vineyard.

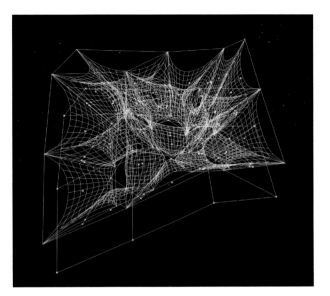
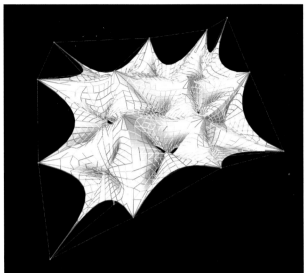
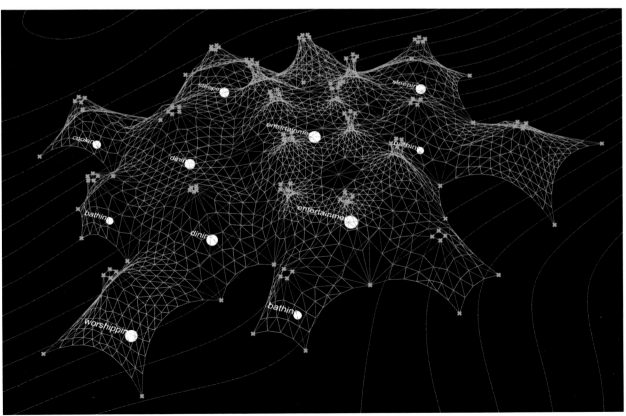

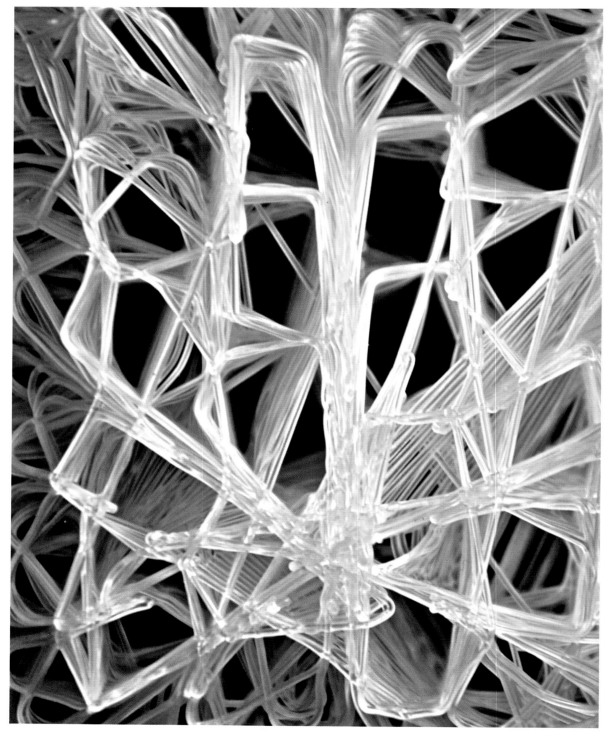

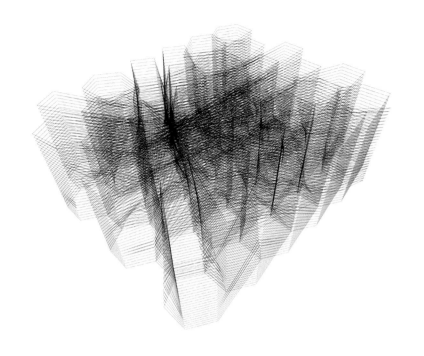

合成纤维模拟 /
Synthetic Felt

将自然感觉的创造和表现作为出发点,这个项目发展了灯丝粗细量级的合成无纺布结构的数字建造设计。在这个结构尺度下进行操作,并利用 3D 打印技术,推动了由 3D 表面或实体模型的新界面的发展。这里使用直接明确的工具路径,而不是被专用软件所决定的工具路径。一项关于控制参数(速度、热、轨迹)的系统化研究提供了操作灯丝粗细和焊接角度的原则。高精度和高等级的产品生产将设计空间延伸到材料尺度。

Taking its point of departure in the fabrication and performance of natural felt, this project develops a design approach to the digital fabrication of synthetic non-woven structures that are specified at the scale of the filament. Operating at this structural scale in the context of 3D printing has necessitated the development of new interfaces which allow direct specification of toolpath rather than toolpaths being determined by proprietary software operating from a 3D surface or solid model. A systematic investigation of control parameters (speed, heat, trajectory) provides insights for steering filament thickness and degree of bonding which extends the design space at the material scale for the production of highly specified and graded artifacts.

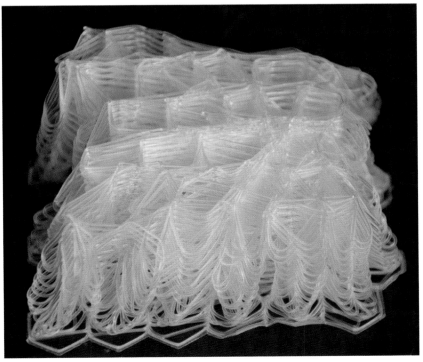

荷兰代尔夫特工业大学 /TU Delft

城市界面 /
Urban Interface

城市界面主要使用了机器人设计与生产的方法,机器人建造团队开发了该方法,并用此方法设计了鹿特丹的介于水面和指定建筑物之间的城市界面。为了探索复杂、变化和从材料的微观层面到建筑空间架构的宏观层面的不同尺度,通过分析当地城市、全球性的社会变革以及 D2RP 制约因素之间的关系,定义了城市界面的应用程序。

The Urban Interface project focused on applying design-to-robotic-production (D2RP) methods developed by the Robotic Building team to the design of an urban interface between the water and the building on the assigned site in Rotterdam. Programmatic use of the urban interface was defined in relationship to local urban analysis, global societal challenges, and D2RP constraints with the aim of exploring hybridity, variation, and componentiality at different scales, ranging from the micro level, as in material systems, to the macro level as in spatial and architectural configurations.

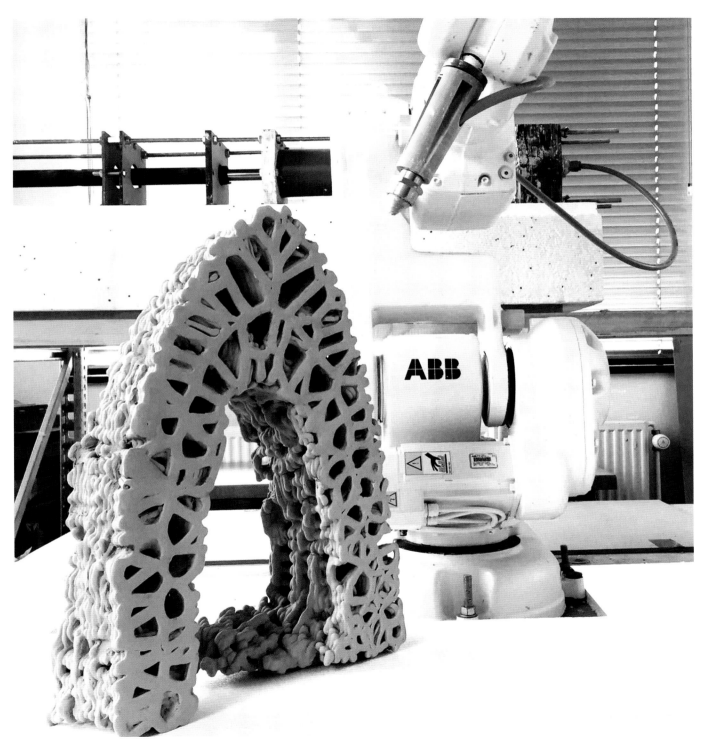

拱 /
Vault

Vault 是使用 3D 建模和对只受压结构进行结构模拟计算而成。它是由经过机械电子切割而成的特殊组件构成。RhinoVAULT 是一款针对只受压结构的交互结构设计工具（由苏黎世联邦理工学院的 BLOCK 研究小组开发）。这款结构设计工具能够完成满足机器人生产条件和施工要求的设计。它的结构是靠手工组装而成，并且为了增加强度，在其表面还覆盖了玻璃纤维和腈纶涂层。

The Vault was designed by employing 3D modelling and structural simulation for compression only structures. It consisted of unique components that were physically produced by employing robotic wire cutting. RhinoVAULT is an interactive structural design tool for compression only structures developed by the BLOCK research group at ETH Zurich. This tool was used to enable the generation of a design that was then informed by robotic production constraints and construction requirements. The structure was assembled manually and was covered with fiberglass and acrylic coating in order to strengthen it.

可伸缩空隙 /
Scalable Porosity

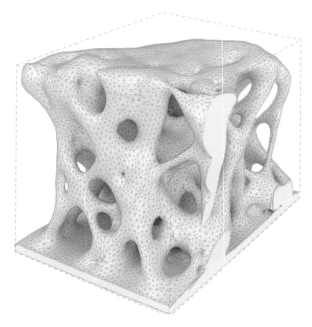

可伸缩孔隙探索了从材料的微观层面到建筑空间架构的宏观层面的不同尺度范围的孔隙。目的是开发出由附加层生成可变孔隙的材料图案。一共进行了两个阶段的实验。一个是纹理研究，另一个是包围体里宏观层面的材料分布研究。这两个初步研究分析了带色黏土的情况，为数控机械材料的沉积提供了输入数据。与早期实验并行的还有如定制配方、测试不同黏度等初步的材料研究。学生们根据个人的需要，自己定制设计—生产的材料信息系统。

Scalable Porosity explored porosities at different scales, ranging from micro levels, as material systems, to macro levels as spatial and architectural configurations. The aim was to develop material patterns that by additive layering generate variable porosities. Two experimentation phases were implemented, one for pattern studies and one to study the macro level material distribution in the bounding volume. These two initial studies later provided inputs for numerically controlled robotic material deposition, in this case coloured clay. In parallel to the initial experimentations through preliminary material studies such as customization of recipes, testing with different viscosities, etc. students customized the materially informed design-to-production system to fit individual needs.

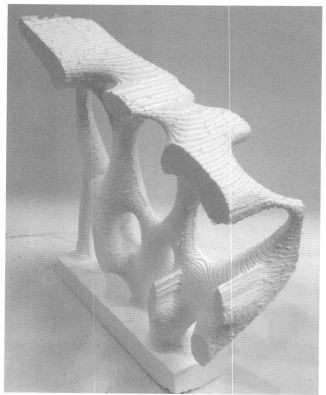

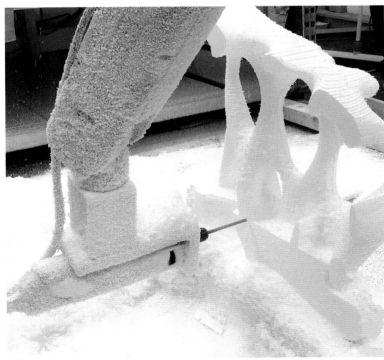

德国德绍建筑学院 /DIA

波塞冬豆荚系统 /
Poseidon Pods

该项目旨在针对阿姆斯特丹市面临的实际问题,在不影响城市文化特征的前提下给出一份灵活的解决方案。该项目的目标是创建一种智能的、适应性强的、可以自行调整,同时又是低碳的结构形式。通过数字设计,映射城市地图,研究不同形式的代理运动方式。我们设计出一种由两部分组成的系统,一部分是可以随阿姆斯特丹水系一起流动的装置单元,另一部分是位于游客较少的区域的装配点。这些装置不工作的时候,还能净化运河的水质。

The project aims to investigate essential and actual problems the city of Amsterdam is facing, coming up with smart solutions that do not affect the city's cultural identity. The goal of the project is to create intelligent, adaptive and self-changing structures, which have a better carbon print on the environment. Using computational design, mapping the city and studying the behaviour of different forms of agents, the design came out as a system composed of units that flow with the water of Amsterdam and assemble points of attraction in areas less populated with people. At the times when they are not working, they clean the water in the canals.

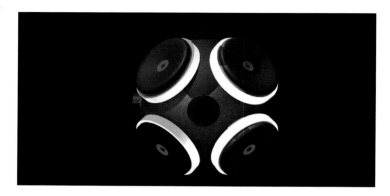

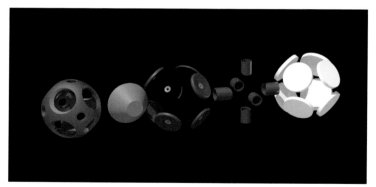

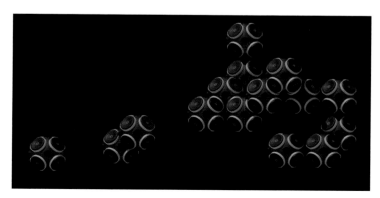

浮体 /
Float Project

浮体项目提出在阿姆斯特丹附近的一个海湾里创建浮动岛屿的概念，这些岛屿可以移动，并且在一定的条件下可以实现自动组织。该项目尝试通过探索不同的组织和行为模式，打破城市现有的按层级划分的树形结构，创建一种半格构状的群系形态。用不同空间类型间的相互作用方式进行设计，是从组成化合物的不同元素分子之间的化学键中得到的灵感。岛屿以海中提取的镁作为其材料，其边缘通过电磁进行控制，使得岛屿之间能够彼此连接或分离。

Float Project is a proposal to create islands in a bay near Amsterdam that float, are mobile and have the ability to self-organize themselves under certain conditions. This project explores different organisational and behaviorial patterns in an attempt to break the hierarchical tree-like design of cities and create semi-lattice formations. Interactions of different spatial typologies are inspired by the formation of chemical compounds that connect with different molecules of different elements. Islands are composed of magnesium extracted from the sea water. The low density of magnesium is well suited for making lightweight structures. The edges are magnetized and charged with electro-magnetic forces which allow the islands to connect and disconnect from each other.

萤火虫系统 /
Firefly System

城市每天都在变得更加复杂，城市问题也变得更加动态化。该项目针对这一问题希望能创建一个新的系统，这个系统能够改变城市固有的自上而下的层次结构，并创建一个新的自下而上的体系。该项目的目标：是让人们在一个城市项目修建之前就可以在视觉上感知到它，能够对其进行测试、与其交互，并得到反馈。该项目由可以实现自动组装的无人机构成，这些无人机有两项任务，其一是要在设计场地上建立的一个全息投影，其二是要创建一个可以供人们行走的物理通道或者路径。这一系统旨在延展空间设计项目在媒体体验与政治活动中的作用。

Cities are becoming more complex every day, and their problems are becoming more dynamic. This project aims to create a new system for cities that hacks its conventional top down hierarchy, and creates a bottom up one. The goal is to allow people to visualize an urban project before it is built, in order to test it, interact with it, and to get feedback. The project consists of drones that can self-assemble, and be assigned two essentials tasks: to create a hologram on site of any proposed design, and to create a physical path where people can walk. The system is intended to extend to embrace media experiences and political movements.

欧洲研究生院 / EGS

人工生态系统 /
Synthetic Ecologies

该项目设想了一种不是供人类居住的栖息地——"自治植物学家"。"自治植物学家"将发展出一种培育人造景观的机器人,这一人造景观亦由机器人完成。这一项目旨在探求一种新型机器人的感性、感知能力的潜能;同时,也为自治系统和美学提出了一种新的思路。该项目尝试转变人们对于机器人代理机能应用的观念,希望能够让观众去拥有机器人的视野、控制机器人的行为,并影响机器人的审美。该项目将会减少人类和机器人在感知能力上的差异,并让人在实体和虚拟方面,从多种尺度和角度进行这种主体客体化的体验。

This project envisions a habitat for non-humanity. Autonomous botanists influence and develop a synthetic landscape nurtured by robots for robots; in essence they are flora-forming. The project seeks to explore the poetic potentials of a new robotic species and puts forward a highly novel approach to autonomous systems and aesthetics. The project seeks to shift perception of robotic agency and allow the audience to embody the robotic gardeners' vision, behavior and influence their aesthetics. The project amplifies perceptual differences between humans and robots and allows for both tangible and virtual embodiment experiences from multiple scales and perspectives.

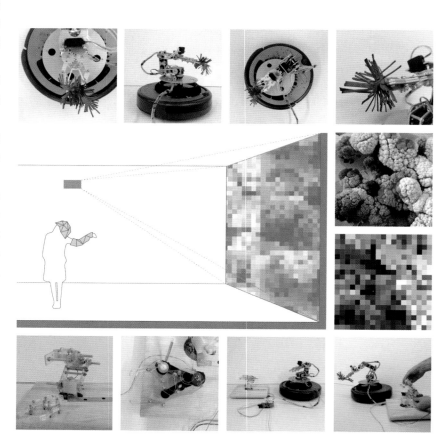

治愈城市组织 /
Healing the Urban Tissue

该项目关注位于埃及开罗市中心的非正式居住区,尝试通过合理地运用算法流程改写现存城市组织的状态。首先,使用"毛绒道路"算法(基于现有的城市肌理)改善城市路网结构,然后对城市现存结构的"生长能力"进行评估,再使用力学向量算法生成新的建筑来增加城市密度,最后使用更深的算法引入城市农业及其他技术,用以增强环境的可持续发展能力。这一项目表明算法不仅可以生成有魅力的新形式,也能够智能地改造现有的城市结构。

This project addresses informal settlements in the inner city area of Cairo, Egypt through the intelligent use of algorithmic processes to rework the existing urban tissue. Firstly, "woolly path" algorithms are used to generate an improved road network based on the existing urban fabric, secondly algorithms are used to evaluate the viability of existing structures, thirdly force vector algorithms are used to increase density by generating more building mass, and fourthly further algorithms are used to introduce urban farming and other techniques aimed at contributing to environmental sustainability. The project demonstrates that algorithms may be employed not just for the production of seductive new form, but also for the intelligent reworking of the existing urban fabric.

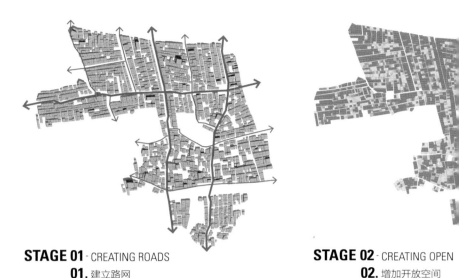

STAGE 01 - CREATING ROADS
01. 建立路网

STAGE 02 - CREATING OPEN
02. 增加开放空间

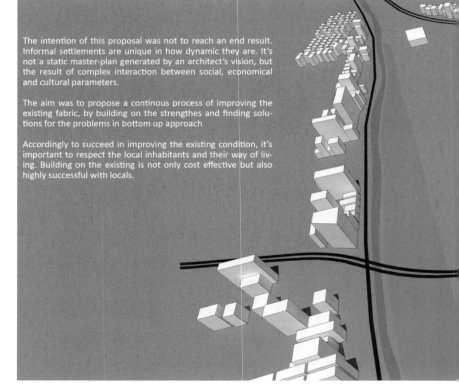

The intention of this proposal was not to reach an end result. Informal settlements are unique in how dynamic they are. It's not a static master-plan generated by an architect's vision, but the result of complex interaction between social, economical and cultural parameters.

The aim was to propose a continous process of improving the existing fabric, by building on the strengthes and finding solutions for the problems in bottom up approach

Accordingly to succeed in improving the existing condition, it's important to respect the local inhabitants and their way of living. Building on the existing is not only cost effective but also highly successful with locals.

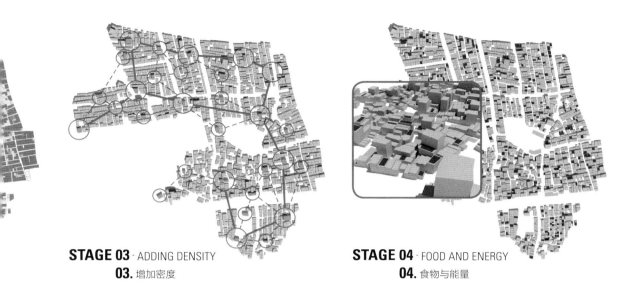

STAGE 03 - ADDING DENSITY
03. 增加密度

STAGE 04 - FOOD AND ENERGY
04. 食物与能量

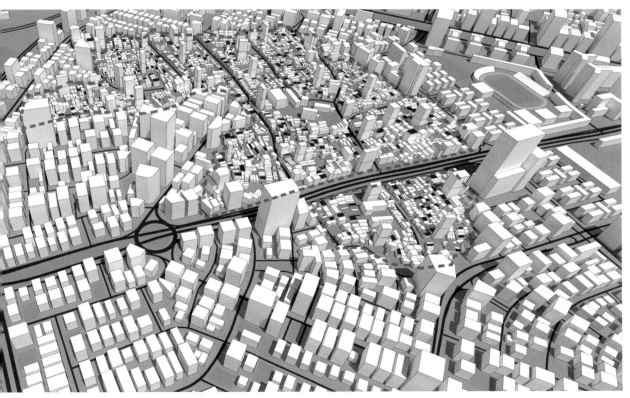

0°N/
0°N

0°N 表明了地球上不久后的未来景象：极地冰雪融化，地球温度上升，世界日益变得不适宜居住。地球上的大多数幸存者将退居两极和青藏高原。为了维持生物多样性的食物链结构，在三个"难民区"之间建立起一种食物交换关系，通过无人机在各聚居区之间运输种子和脱水食物。通过算法和网格松弛方法生成居住设施的抛物线表面，以增大接受光照的面积，同时通过有追踪技术的太阳能镶板，最大程度地为温室收集太阳能。通过一套集群形式算法，实现不同构筑群之间的差异，形成一个有反应的、多样性的景观。

0°N suggests a near future event where the polar ice caps have melted, planetary temperatures has risen, and the world is largely inhospitable. The majority of the Earth's surviving population has retreated to the two poles and the Tibetan Plateau. To sustain a bio-diverse diet, three refugee zones are engaged in a complex food exchange where drones transport seeds and dehydrated food between populations. Algorithmic form generation and mesh relaxation sculpt parabolic surface curvature to increase solar reception in the living facilities, while object-driven panelization techniques maximize solar collection for the greenhouses. Diversity between constructs is enabled through an algorithmic population of form, enabling a responsive, diverse landscape.

西班牙加泰罗尼亚高级建筑研究学院 /IAAC

KATA /
KATA

KATA 项目直接回应了生活在蒙古的游牧民族所面临的沼气排放、由气候引起的荒凉环境、资源稀缺等诸多紧迫的问题。冰冻的生物深层土，也就是永久冻土，在持续不断地向大气中释放出大量的沼气，给我们带来危险。如果立即着手解决，这个问题可能会得到缓解，长久看来这会成为一种用之不竭的可再生能源。随着技术的进步和持续不断的研究投入，沼气资源将慢慢地成为蒙古周边富裕生活的推动力。

The Kata project is a direct response to pressing issues of Methane Release, harshening climatic conditions and scarce life resources for nomads in Mongolia. A deep layer of frozen biomass, permafrost, has been releasing huge quantities of methane gas into the atmosphere, putting our population in danger. If addressed immediately, this problem might be alleviated and furthermore, turned into a source of seemingly endless source of renewable energy. With advances in technologies and continuous input from the world intelligence methane gas will slowly become a driving force for prosperous life inside and outside of the Mongolian borders.

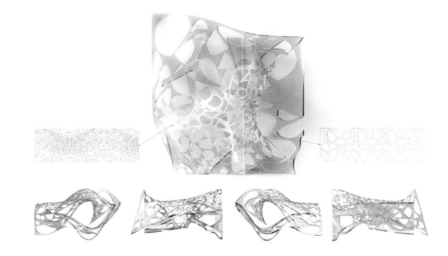

转译的几何 /
Translated Geometries

为了将这个项目应用于可动的建筑原型，转译的几何和形状记忆聚合物一起使用。建筑在形态变化的过程中，使用了一种材料，这种材料在受到外部可控的刺激之后会改变相位。材料用在完全嵌合的三角折纸原型的几何形态系统中的结构连接点上，形状记忆聚合物的使用量被最小化。激光切割成型的聚合物节点被置于这个几何形态的最高点和最低点的交叉点，以通过这些节点的变化来控制整体的形态变化。从起初整体结构中较为平滑的位置开始，这些位置被全部加热、变软，然后经过无人机的驱动作用，重新达到希望的位置。

Translated Geometries works with Shape Memory Polymer (SMP) in order to apply it to a responsive architectural prototype, architecture in transition, using a material that can change phase from an external and controlled stimuli. The use of SMP is minimized to perform as a structural joint in the material system of the geometry, that of a tessellated triangulated origami pattern. The laser-cut hexagonal SMP joints are placed at the intersection of the pattern's mountains and valleys to control the overall geometrical deformations from these individual nodes. From the initial flat position of the homogenous structure it can be entirely heated and softened and then deformed into a desired position through the actuation force of pulling by many octocopter drones.

通达的潜力 /
TRANS[X]PONENTIAL

"通达的潜力"批判性、前瞻性地质疑当前的城市运输能力。这个地段位于巴塞罗那一个环路层叠的地方，尽管它是城市间交通连接的枢纽，但它降低了居住环境质量和土地价值。这个方案将城市生活从按照函数式快速增长的公共基础设施建设中释放出来。作为运输的车辆单元将不再依附于道路网络，成为一个自主飞行器：具备传感器，基于二进制进行数据交互，并使用太阳能。这是对城市的变革，技术在建筑、交通和用户等多个层面发生了改变：建筑变得适应空中汽车，且交通管理变得实时化。这个综合系统可以使塞尔达的交通堵塞情况从本质上转变为随机编码的形式。

Trans[x]ponetial interrogates current urban mobility with a critical forethought to upcoming conveyance models. The site in Barcelona is a layered overlap of ring roads. Though it connects at intercity level, it declines quality space in terms dwelling environment and economy of land. The proposed system liberates urbanism from rapid infrastructure growth as a function of increasing program. The vehicular unit of mobility is no longer adhered to road networks becoming an autonomous aerial vehicle: sensor enabled, data interactive body driven by binary and solar power. This revolutionizes the city. Technology shifts at building, traffic and user level: buildings mutate to accommodate the skycar and traffic management is real time. This comprehensive system shall metamorphose the gridlock of Cerda to a random codedness.

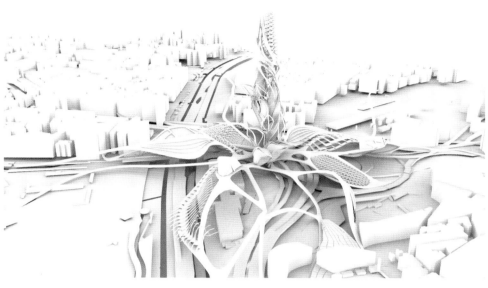
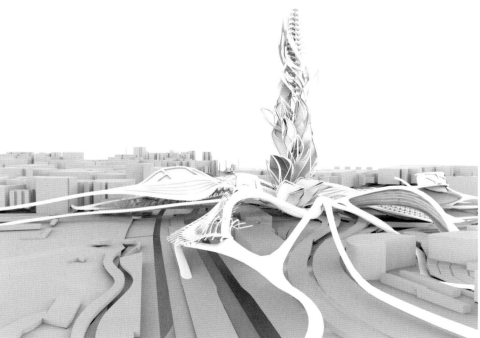

法国巴黎玛莱柯建筑学院 /Paris Malaquais

自由度 /
A Degree of Freedom

该项目旨在研究物质、模拟、控制和数字制造。为了嵌入可变材料，我们很快聚焦于智能材料，调研软框架电活性聚合物在动态建筑领域的潜力。20世纪90年代，材料科学家已经将其用于传感器、驱动器和人造肌肉。为了获得准确和精良的制作方法和工具，这个项目致力于掌握其制造过程，发现新的组装方法，发展模拟和控制系统，努力拓展其在建筑中的应用。

This project aimed at being applied research focused on materiality, simulation, control and digital fabrication. With the clear will to get a dynamism embedded within the studied material itself, we quickly refocused our topic on smart materials and decided to investigate the possibilities of Soft Frame Electroactive Polymers (SFEAPs) in the kinetic architectural field. Materials scientists and engineers have been developing electroactive polymers for use as sensors, actuators, and artificial muscles. Necessitating very accurate and sophisticated production methods and tools, this project was dedicated to mastering the fabrication process of SFEAPs to the experimentation of new assembly approaches, and to the development of a simulation and control system, aiming at developing an architectural application of SFEAPs.

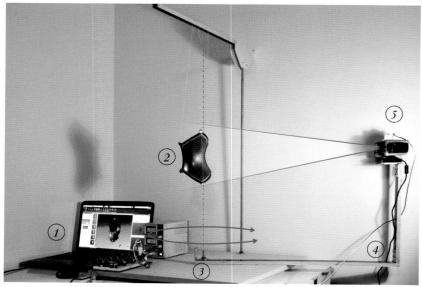

a degree of freeform   kinetic structure

60

机械臂堆砌 /
RO-BORIE

本研究旨在展示计算机科学，尤其是机器人在建筑生成中的影响。案例是干石屋，这种法国南部典型的建筑方式在先进的技术中被重新诠释。这个项目考虑到特定的文化背景，借鉴传统架构，利用机器人获得了解决方案。如果一个工匠能够建造复杂的形体，那么这次搭建则拥有必备的精度、稳健性和快速性。其意义不仅是建筑能力的拓展，还是人与机器人之间共生的真正协同作用。

This research aims to demonstrate the impact of computational sciences, and particularly the use of robotics, in the updating of architectural production. The case studied is that of a dry-stone hut, or borie. This dry-stone construction, typical in the south of France, is reinterpreted through advanced techniques. The project is not developed from scratch. It takes into account a particular cultural context and draws on the teachings of traditional architecture to allow the emergence of experimental solutions, provided by robotics. If a craftsman is able to shape complex elements, the performances explored here suggest an essential precision, robustness and rapidity. Beyond an extension of the architectural capacities, the symbiosis between man and robotics allows a real synergy.

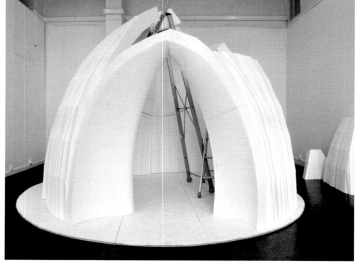

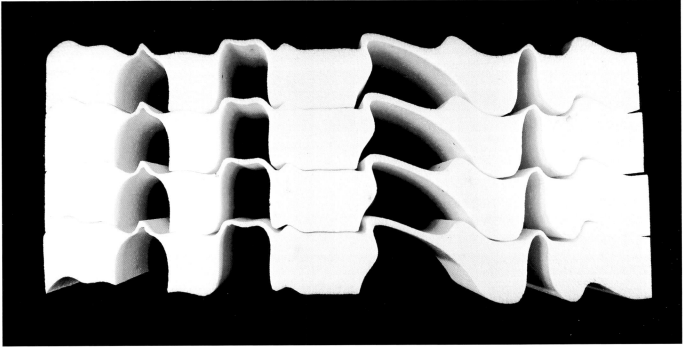

加法建造 /
Additive Manufacturing

加法建造不仅具备解决将结构分解为"原子"元素的问题，而且也破坏了传统的建造序列。事实上，源于计算机的控制，我们可以设想复杂的形式，这种复杂不仅体现在建筑的形成，还存在于建造策略之间的对话。加法建造能够根据材料特性创造最佳的几何形式；能够以简单的方式建造更加复杂的几何形体；能够以精确的方式综合考虑材料、多样性能、技术和资金成本等问题。

Additive manufacturing not only poses the question of the breakdown into 'atomic' elements of the structure, but it also disturbs the traditional sequences of construction. Indeed, we may envisage the impression, in situ, of complex forms since they are controlled by computers, not only in the total formation of the building, but in dialogue with manufacturing strategies. We believe the additive production is relevant, amongst other things, to control the geometry to distribute the different materials in an optimal fashion depending on their performance; to implement reservations, in a simple fashion, even in complex geometries destined to welcome another constructive element; and to combine, in a very relevant and precise manner, materials with very different performances and technical and financial costs.

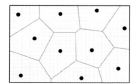 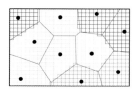 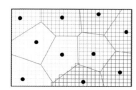
 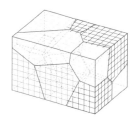 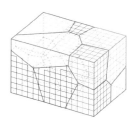
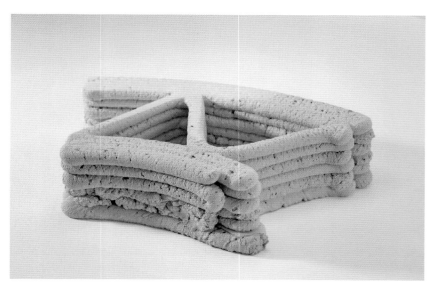

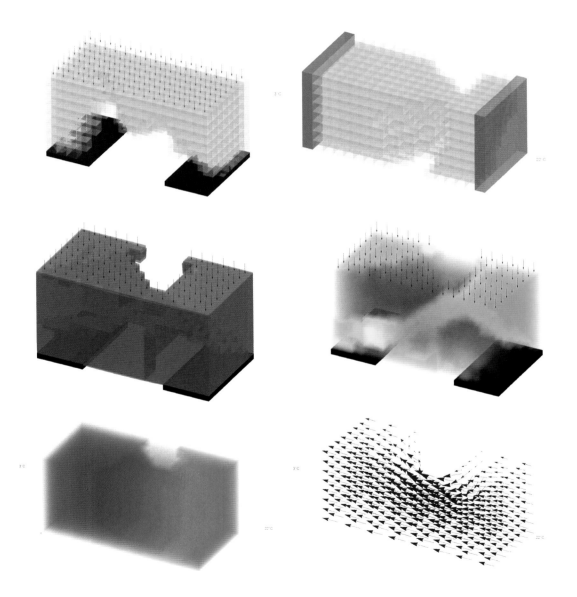

德国斯图加特大学 /ICD Stuttgart

可扩展表皮 /
Augmented Grain

可扩展的表皮研究在增强与数字映射、选择、加工、装配,以及天然材料系统相关的如木材吸湿特性方面的能力。将低技术的传统木材加工工艺和先进的计算制造相结合,可以针对每一块复合木材表皮的方向性、强度以及膨胀收缩度进行功能性多元化编码。此研究意味着建造技术方法的转变,即从设计、制造及手工组装几何形状的静态元素,转变为设计出能够应对环境条件变化,进行编程和自我塑形的活性表皮。

Augmented grain investigates the potential of harnessing and enhancing the existing hygroscopic properties of wood in regards to digital mapping, selection, fabrication, and assemblage or natural material systems. A combination of low tech traditional woodworking techniques and advanced computational fabrication methods are used to encode functional variation in relation to orientation, intensity, and speed of expansion and contraction within each composite timber surface. The research project represents a methodological shift in construction technique from designing and fabricating static elements that are manually assembled for a specific geometry, to designing active surfaces programmed to self-form in response to changes in environmental conditions.

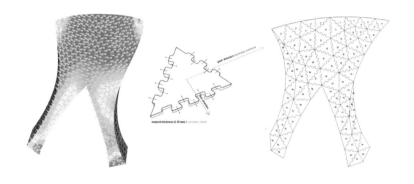

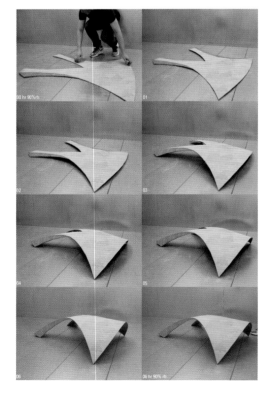

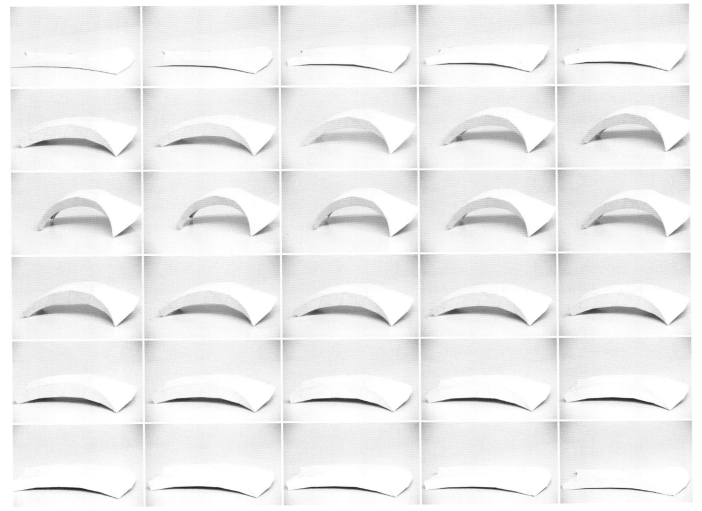

编码表皮 /
Encoded Surfaces

该项目探索了一种通过定义其网格的拓扑结构来设计综合性双曲面编织物表面的方法。受到非欧氏几何体系的启发，利用这种不同的技术解决此类编织结构的非常规展示、模拟及制造问题。纤维编织图案清晰的拓扑关系可以通过几何镶嵌实现抽象编织网络。通过增减二维图像特定区域的顶点以及多边形数量，均衡节点间距，产生三维变形量，这样便于利用分化高斯曲线来实现几何生成的定制要求。

The project explores a generative approach to designing complex doubly curved woven textile surfaces of different integrated materiality by defining the topological pattern of their thread networks. Different techniques are examined for solving the non-trivial representation, simulation and fabrication problems related to such woven structures. The clear topological relationships within the weaving pattern of the fibers enable an abstract representation of the textile network as a geometric tessellation. It can be customized to produce geometries with differentiated local Gaussian curvature - this happens by either increasing or decreasing the amount of vertices or polygonal faces in specific areas of a two-dimensional diagram and subsequently equalizing the node distances to produce three-dimensional deformation.

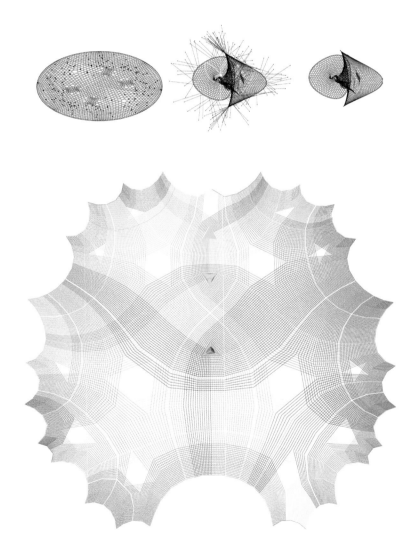

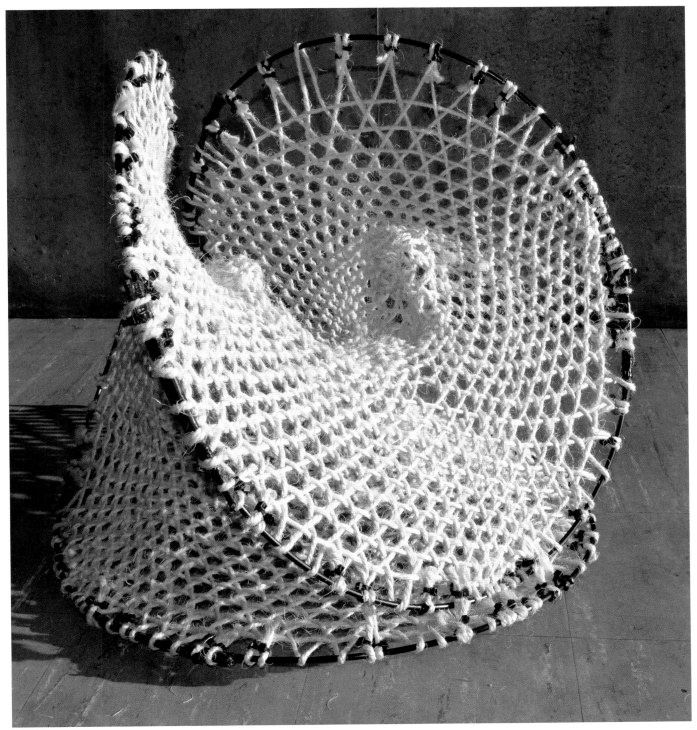

碳纤维亭子 /
Carbon Fiber Pavilion

这个项目的重点是一种平行的自下而上的设计方法，这种设计方法针对天然复合材料的贝壳状曲面进行仿生研究，并且对于增强纤维聚合结构的虚拟机器人的制作方法进行了发展。这个项目的目标是发展模块化曲面技术，是一种双层纤维复合结构，这种结构可以用最少的模板做出最复杂的曲面结构。为了制造独特几何构造的纤维，开发出双曲模块机器人空心绕组方式，这种方式采用了两个六轴工业机器人在两个既定形态的钢架之间进行纤维缠绕。

The focus of the project is a parallel bottom-up design strategy for the biomimetic investigation of natural fiber composite shells and the development of novel robotic fabrication methods for fiber reinforced polymer structures. The aim was the development of a winding technique for modular, double layered fiber composite structures, which reduces the required formwork to a minimum while maintaining a large degree of geometric freedom. For the fabrication of the geometrically unique, double curved modules a robotic coreless winding method was developed, which uses two collaborating 6-axis industrial robots to wind fibers between two custom-made steel frame effectors held by the robots.

瑞士苏黎世联邦理工大学建筑学院 /ETH Zurich

连续框架 /
Sequential Frames

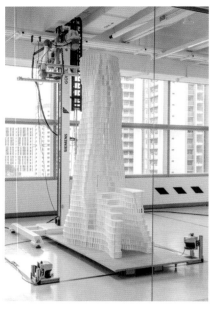

该项目定义了一个高层内部空间，由计算机编程控制密集放置的剪力墙，构成系列空间。计算设计并确定了这些砌成的墙，计算时考虑了从上到下的力的传递，以及高层塔楼被使用者所期望的混合功能。5200面墙由纸带制成，这些纸带先以特殊角度放置于真空夹中，再由机器人切割而成。接着机器人再进行翻折、涂胶并置于模型上。只有这种完全集成的计算设计和机器人制造方法才能完成高度复杂的室内空间设计。

This project defines the interior spaces of a high-rise not by enclosure, but by computationally programmed cut-outs in a sequence of densely placed shear walls. Here, a computational design process defines these cut-outs by negotiating between force flows, calculated from top to bottom, and the desired architectural typologies for the mixed-use programme of the tower. The 5,200 wall elements are made from paper strips that are robotically cut after having been placed into a vacuum clamp at specific angles. The robot then folds the pieces, applies glue and places them onto the model. This approach towards fully integrated computational design and robotic fabrication routines allows highly articulated designs with a large variety of interior spaces.

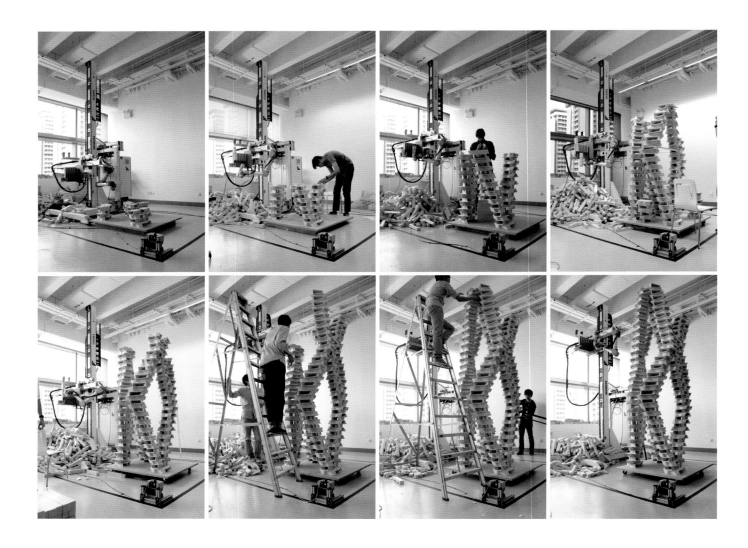

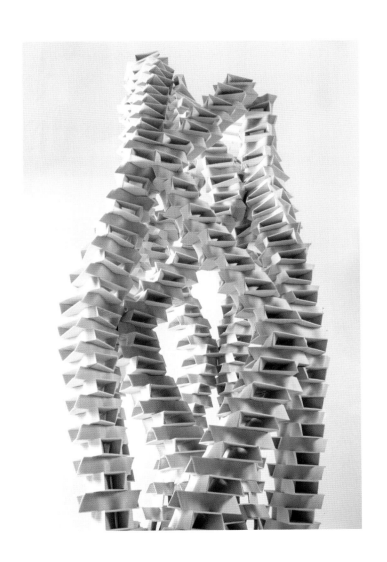

网格塔 /
Mesh Towers

网格塔的建造基础是计算设计和机器人装配成的几个细长塔楼，它们在生成高度的过程中合并、分离，并在结构上相互支持。每座塔由背离其直接相邻单元的聚苯乙烯泡沫塑料单元构成。这些单元由几何异化的结构要素构成，利用机器人精确移动物块的能力，机器人拿起一个物块，沿计算生成的路径来移动并通过热丝将其制造出几何形状复杂的墙面，为装配做好准备。错综复杂的墙面及单元的设计遵循结构的引导曲线，而曲线是实现整个塔楼的连续荷载的传递路径。

Mesh Towers is based on the computational design and robotic assembly of several slender towers, which merge and separate as they grow in height, structurally supporting each other. Each tower strand is constructed out of stacked Styrofoam units that are oriented away from their immediate neighbouring elements. These units are constructed from geometrically differentiated structural elements, taking advantage of the robot's capacity to accurately move a block: the robot picks up one element and moves it through a hot wire along a computationally generated path to fabricate a geometrically complex wall element, ready for assembly. The intricate wall (and unit) designs follow structural guiding curves to achieve continuous load transfers throughout the whole tower.

垂直大道 /
Vertical Avenue

垂直大道提出了一个螺旋式循环坡道系统，该系统由一系列不同的公共项目垂直分布在塔楼中。计算设计的螺旋坡道系统连接几个含有住宅单元集群的结构核心。传感器集成落足装配组件的高精度需求，使无缝集成的人工流程补充机器人操作成为可能。此外，该项目利用数字化控制的受热变形材料和外立面元素装配而成。该设计和制造的概念确保了一个强大、开放的机器人组装方式，将大量建筑单元实物砌筑成不同的高层结构体。

Vertical Avenue proposes a spiralling circulation system with a range of different public programmes distributed vertically throughout the tower. The computationally designed spiralling ramp system connects several structural cores where residential units are arranged in clusters. The integration of sensors allowed for both matching the needed precision in the placement of components as well as the seamless integration of manual processes complementing the robotic manipulation. Additionally, the project makes use of digitally controlled thermal deformation and assembly of facade elements. This design and fabrication concept ensures a robust yet open approach towards robotic assembly, allowing the physical manipulation of a multitude of building elements into a differentiated high-rise structure.

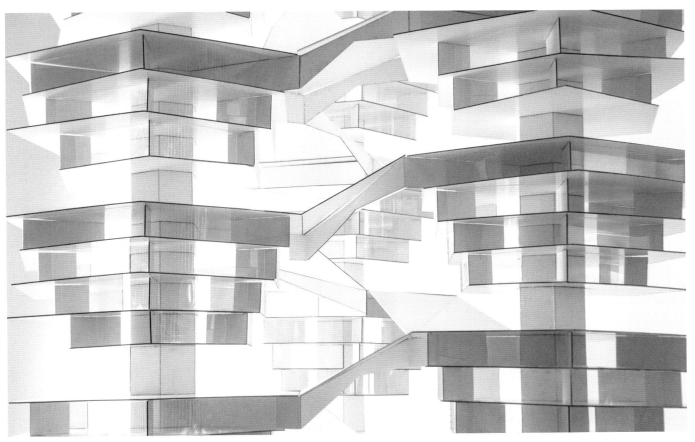
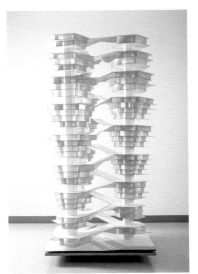

澳大利亚皇家墨尔本理工大学 / RMIT

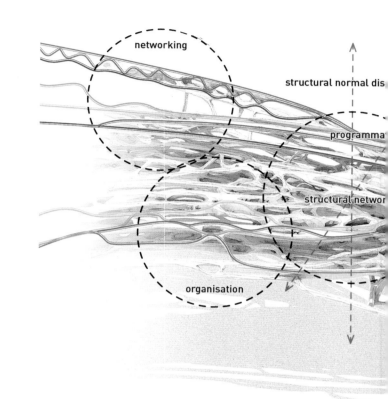

微丝 /
Ibrosity

这个项目尝试使用算法的方式将设计师的意图注入项目中。它通过创造解决真实建筑问题的定制行为来挑战简单应用已有算法的常态。它试图强调的是建筑算法和应用算法之间的区别，以智能算法为基础，这些算法通常用来模拟鱼群和鸟群。该算法执行和响应 FEA 分析，尝试在局部改变它的形态以获得较优的结构特性。

This project attempts to use algorithmic means to imbue the intent of the designer into the project. It attempts to challenge the normality of simply applying existing algorithms to architecture through the creation of customised behaviours that address real architectural problems. It is, therefore, an attempt to highlight the difference between an architectural algorithm versus an applied algorithm. It begins with an agent-based algorithm, those generally used to simulate things such as schools of fish or flocks of birds. The algorithm performs and responds to FEA analysis, locally attempting to alter its form to allow greater structural performance.

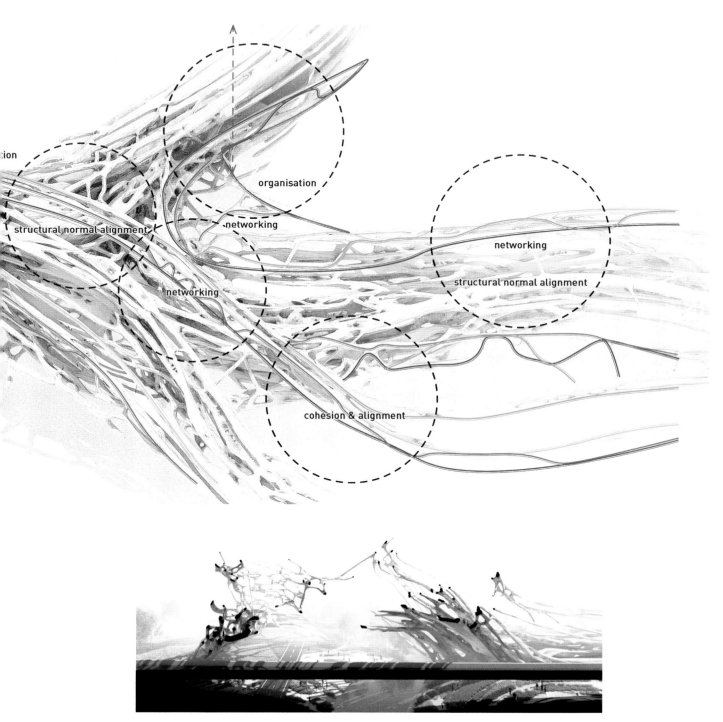

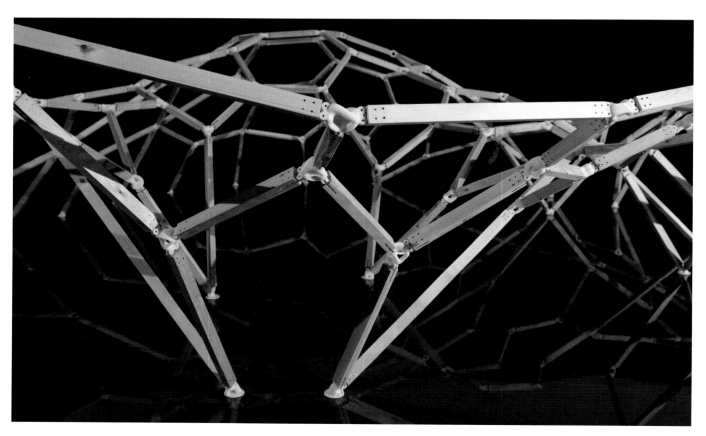

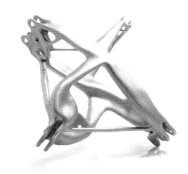

智能节点亭 /
SmartNodes Pavilion

该研究项目探索了用特殊配对、尖端科技、现成组件的重量最优节点、标准梁、固件等完成定制的复杂建筑形式的可能。它解决了设计和制作轻量管层结构原型的难题。该系统以一系列制造定制结构的节点为基础，用数字算法来设计和优化，并使用叠加式制造技术进行制作。它们可连接简支梁和平面面板。通过对每个节点几何形状的定制，使该系统具有不用花费大量成本便能产生各种设计结果的潜力。

The research project explores the potential for pairing unique, high-tech, weight optimal node components with off-the-shelf, standard beams and fixings to accomplish customised and complex building form. It addresses the challenge of designing and prototyping a lightweight canopy structure. The system is underpinned by a series of custom manufactured structural nodes, designed and optimised with digital algorithms and produced using the latest additive manufacturing technology. These link simple beams and planar panels. Through the customised geometry of each node, the system has the potential for a broad range of design outcomes without significant cost penalties.

空间场 /
The Spatial Field

该项目以塔楼类型为例,以过程为基础探讨了建筑物的建立和发展,与多个进程一道具有正式且程序化的特征,鼓励不同程序间的交互和联系。选择和使用的策略是冰川形成的进程。通过对其行为的广泛研究和阅读,我们创造了一个可变格点的系统,同时也考虑到地段的关系和连接。另外还运用几何学来探索特定的影响,试图通过创造形式来建立令人敬畏的空间。这里,过程及结果的评判是非常必要的。

This project explores an architecture established and evolved through a procedural based exploration within a tower typology, working with multiple processes to achieve the formal and programmatic qualities that encouraged interweaving and social density between programs. The device chosen and used is the Glacier Formation process. Through extensive research and reading about their behaviour, a system of variable grids was created and applied to the site, whilst site relationships and connections were also considered. A parallel interest was also working with geometry in search for a particular affect, which was trying to create spaces of awe through form making where the process and a later critique of the outcome become imperative.

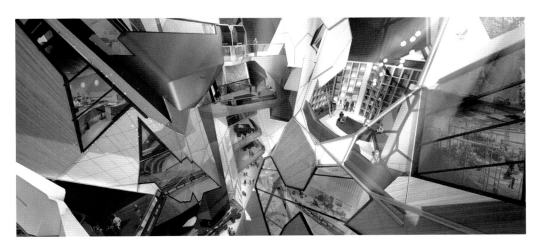

中国香港中文大学建筑学院 / CUHK

香港文化中心 /
HK Cultural Centre

香港文化中心设计项目，体现了设计任务书"高度唯物主义"的要求，探索了"程序空间设计技术"及"软件空间生成能力"并用于建筑设计，包括连续围合几何形体、完整的建构和高度可感知的材料表达。非欧形式用作一种空间语言来回应密集性、互联性和围合性几个方面的主要需求。一个鹅卵石状覆盖系统夸大了这些几何形态。这些图案随着表皮的走向和流动，形成整体化的装饰效果。最终结果试图激发使用者感觉及触觉上的反应。

This design for a Hong Kong Cultural Centre responds to the studio brief "Hyper Materialism" by exploring how procedural spatial design techniques and recent software's space-generating abilities allow for the design of architecture that embraces continuous geometries, integrated tectonics and a highly tactile material expression. Non-Euclidian form is applied as a spatial articulator responding to programmatic requirements in terms of density, interconnectivity, and enclosure. A shingle cladding system exaggerates these spatial geometries. Patterns follow the orientation and flow of the underlying surfaces creating an integrated ornamentation. The final outcome seeks to provoke sensual, haptic reactions from users through catalytic architecture.

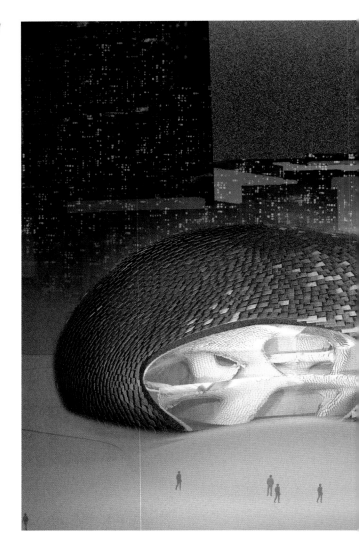

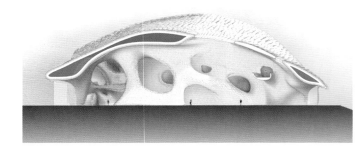

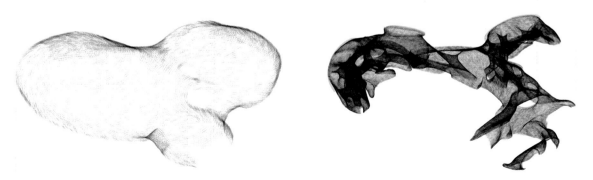

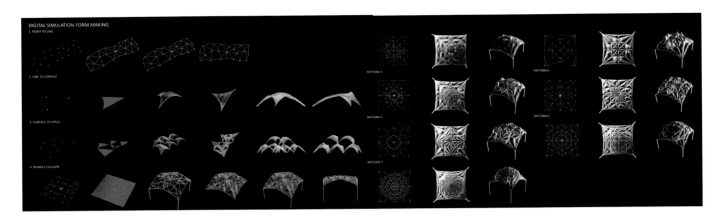

哥特式重建 /
Gothic Reconstruction

这个项目通过实时物理引擎作为找形工具，发展和应用哥特式建筑结构及空间体系，以此来满足设计任务书要求的"高度唯物主义"。该项目开发出一个集合建构系统，它把表皮生形工具程序与数字反转悬索几何设置相结合，从而定义了一系列的高装饰性的拱形空间。这些拱形空间用于香港文化中心的大厅区域，并且将穿越建筑的多股人流组织在一起。高度错综复杂的天花板被放大了，作为表现殿堂文化的建筑装饰功能。

This project responds to the studio brief "Hyper Materialism" by developing and applying Gothic architecture's structural and spatial systems through the introduction of real-time physics engines as form-finding tools. An integrated tectonic system is developed that combines procedural surface patterning tools with digital inverted catenary geometry setups in order to define sequences of highly ornate vaulted spaces. These vaulted spaces are deployed in the lobby area of a Cultural Centre in Hong Kong and weave together the multiple flows of people that navigate throughout the building. The highly intricate complexity of the ceiling amplifies the building's function as a temple for cultural expression.

竹网架 /
Bamboo Grid Shells

香港传统的竹脚手架搭建技术濒临消失，如何利用数字生形工具将这一技术应用于轻质的、具有弯曲弹性的竹网架结构建筑的设计与建造是本项目的研究内容。我们将于 2015 年在香港建造一个大型的临时竹网架展棚。方案来自于香港中文大学建筑学院的学生工作营，并由其设计团队与结构工程师进行合作。本项目一方面为大跨度建筑寻找可持续性的、轻质的构造方式，另一方面也为传统技术寻找在 21 世纪的发展出路。

This research project showcases how the endangered craftsmanship of Bamboo Scaffolding Construction in Hong Kong can be expanded with the architectural design and construction of light-weight, bending-active, bamboo grid shell structures through the introduction of digital form-finding tools. A large temporary bamboo pavilion is being built in Hong Kong in 2015. Its design is the outcome of a Student Design Workshop held at the CUHK School of Architecture, and was developed further by a research team in collaboration with structural engineers. By promoting sustainable, lightweight construction methods for large span architecture the project seeks ways for traditional craftsmanship to evolve for the 21st century.

中国香港大学建筑学院 / HKU

媒体结构 / Media Tecture

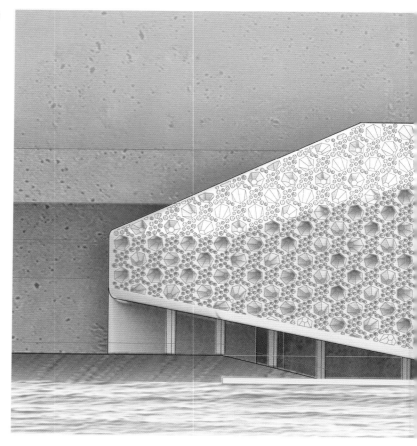

本毕业设计通过研究媒体外墙的潜力，力求创造如同身临其境般尺度的公共空间。与建筑相关的是持久性和深度性，而媒体表现则被视为短暂性和表层性，我们可以认识到这两者存在对立及由此失去的潜力。媒体式立面往往被开发商或是政府部门所喜闻乐见，但建筑师通常对此表示不屑。本设计研究和探讨了光技术的发展，以求在深度、阴影、夜间光照等方面提高媒体式建筑立面形式的品质。

This thesis project seeks to explore the potential of media facades by investigating light and architecture to create immersive scalar public spaces. Architecture and media have had a contentious relationship so far. Whereas media is seen as temporal and superficial, architecture is associated with permanence and depth. Today, we recognize the struggle and lost potential of the relationship. Media facades are loved by developers, or city mayors, but are often loathed by architects. This thesis studies and investigates the development of light technology in order to enhance architectural qualities of the media facade in form, depth, shadow, and mass at night through media and light.

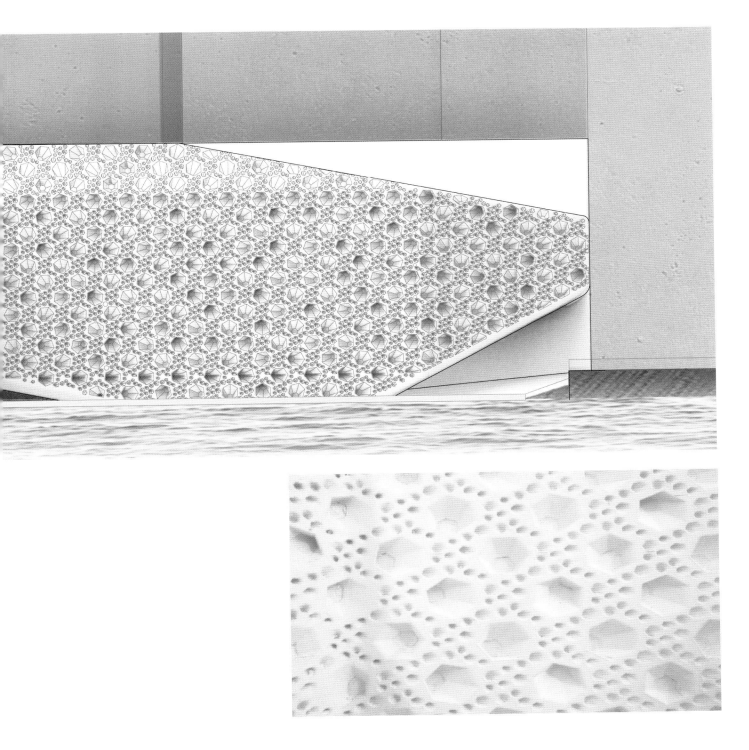

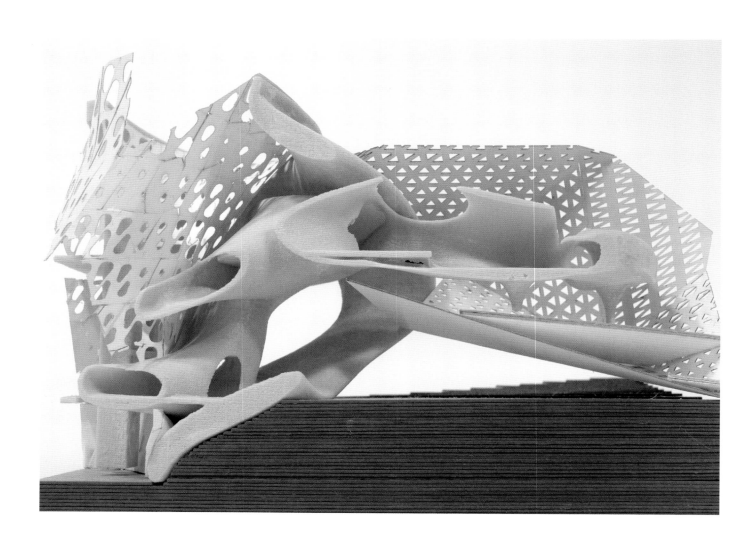

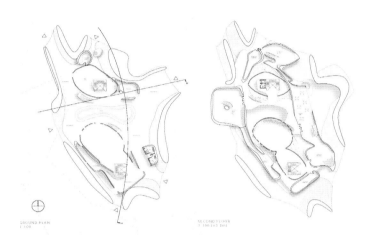

博物馆热潮 /
Museum Boom

中国城市的博物馆建设热潮使得博物馆在城市中的内涵发生了变化，从单独的建筑设计到建筑规划中"文化区域"的划定，都可以看出这种现象的端倪。设计的方法着重于设计生成展开过程的清晰模式，这一模式面向一系列相似的城市和建筑系统以及相关性不强的内陆传统文化建筑，同时也面向国外文化范畴。该项目通过发现激进的建筑表现所带来的影响，来面对当代建筑文化下的装饰语境。

The urban implications of the museum building boom in China may be found in the shift from the isolated phenomenon of individuated cultural buildings to design initiatives which are coordinated and designed as "cultural districts", thus forming the brief for the studio, for a site in Chengdu. The design research methodology focused on generating and deploying discernible and legible patterns, towards a set of correlated architectural and urban and architectural systems, and the often disjunctive relation of vast interior urbanism of cultural buildings, and their exterior civic territories. The studio confronted the recapitulation of the ornamental in contemporary architectural culture, through searching for unfamiliar affects of a radical expressive architecture.

系统的多重性 /
Systems of Multiplicity

在香港建筑高密度条件下的过去二十年里，高效经济的塔楼成了主要的建筑模式，这一模式已成为我国当代城市住宅建设的一个参考。在这个项目中提出的关键问题是：对于高密度的居住环境，此种大量聚集的建筑类型是否是正确的？如果不正确，那么还有其他合适的类型么？我们试图寻找问题的答案，并进行了大量的实验和替代方案的研究。这些替代方案是运用新的数字设计技术产生的差异、变化和特异的结果。

In the past two decades the podium tower typology has emerged as the predominant building model of efficiency and economy for high-density housing in Hong Kong. Now this model has become a reference for contemporary urban housing developments in China. The key questions asked in this studio are: is this typology aggregated in huge numbers the right model for high density living, and if not what are the other models that could be suitable? The studio tried to find answers and invested heavily in the experimentation and production of alternative solutions by establishing new digital design techniques that can produce difference, variation and specificity.

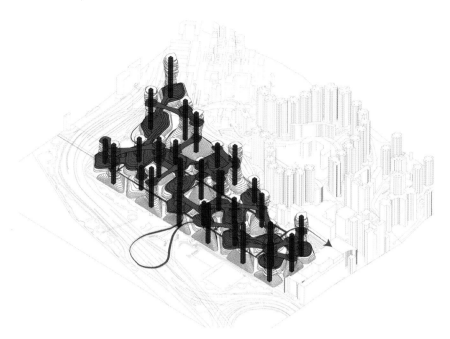

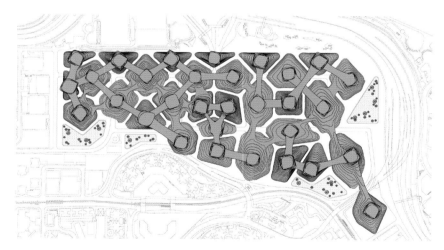

中国南京艺术学院 /Nanjing UA

无器官身体 /
Body Without Organs

无器官身体是身体的虚拟延展部分。一个实体的有机体是由一个空间区间内的基本粒子和力构成的。通过定义程序和运动方式，产生一个新的"身体部分"，并通过特定的作用和程序，使其成为人原本的身体与其复杂环境的新的联系。该项目旨在探索无器官身体在这方面的潜力。

The body without organs [BwO] is a concept taken from the French philosopher, Gilles Deleuze. It refers to the virtual field of the body, the domain of the basic particles and forces ("singularities," "affects," "intensities," "ideas," "perceptions," and so on) from which an actual organism is composed. The aim of the BwO workshop is to explore the potential of the BwO, as a way of reconnecting the body with its complex environment with specific affects and programs, from which a 'new body' is generated from its defined programs and movements.

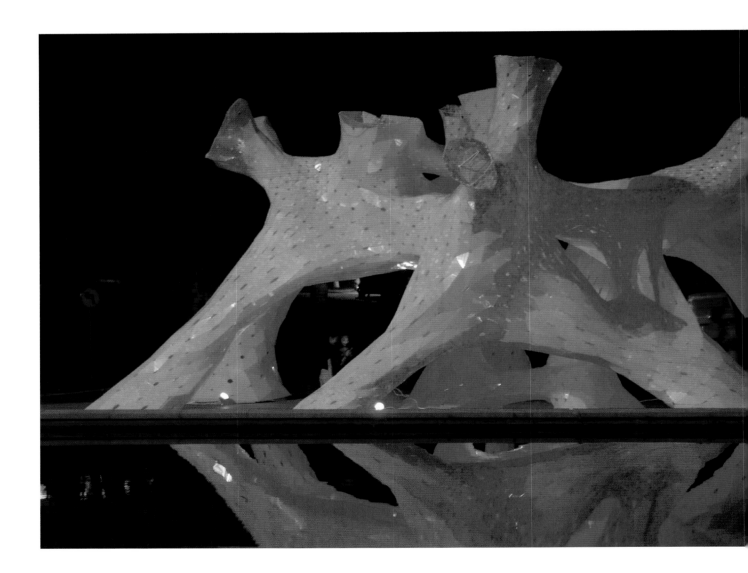

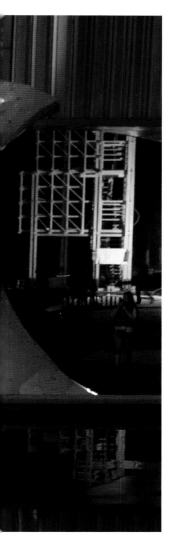

数字森林 /
Digital Forest

数字设计的发展起源于对于自然系统复杂性的研究。榕树根系形成的空间是我们本次研究的切入点。这是一种复杂的、随机的、自由的同时也是不确定的空间。我们希望来模拟和再现自然生长出来的空间，然后利用可回收材料建造这个空间。

The development of the design for the Digital Forest originated from research into the complexity to be found in natural systems. The research started by looking at the space that is constituted by the banyan tree root system. This allowed us let us find the breakthrough point in our research. This space was not only intricate, random and free but also uncertain. The intention is to simulate and reproduce the space of a natural growth system, and then to use recyclable materials to construct the space.

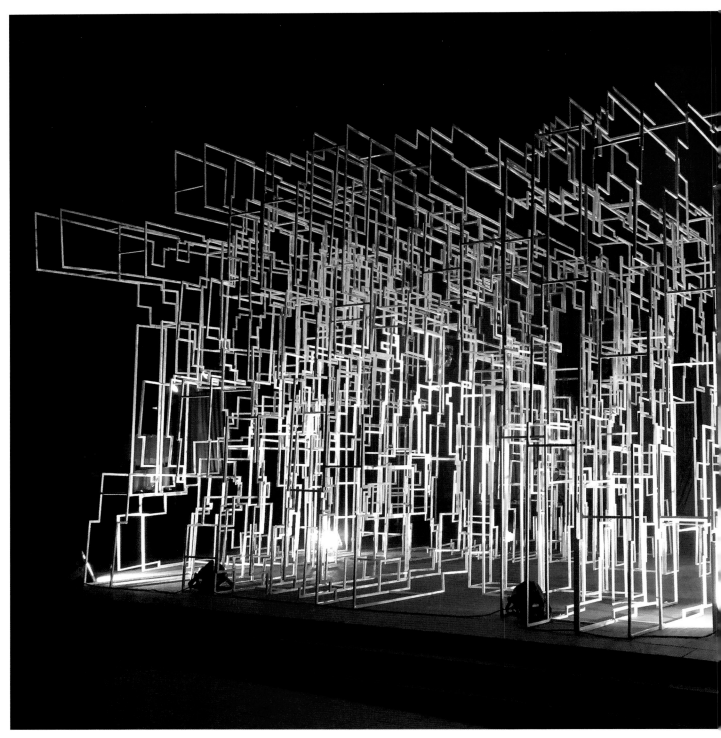

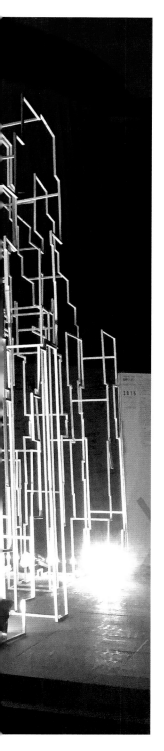

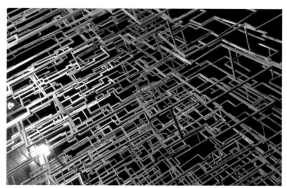

多维假山 /
Multidimensional Rockery

我们从中国传统景观文化中提取元素，分析各种自然形式的假山产生的复杂机制，重点讨论参数化的找形和生成方法，利用参数化设计技术来设置相应的生成逻辑和形态方式。最终我们用抽象线条和多维空间的形式来表达这一设计。

We started off by extracting elements from the Chinese traditional landscape garden. We then analyzed the various natural forms of the rockery's complex generation mechanism. Next, focusing on various discussions about parametric design methods for form-finding and form generation, we utilized parametric design technologies to set up a corresponding generation logic and to materialize the form. Finally the design is expressed through abstract lines and presented as a multidimensional spatial form.

中国华南理工大学建筑学院 /SCUT

"领"椅 /
Collar Chair

该设计以衣领作为形态原形,为了表现它的优雅气质,选用不锈钢作为材料,以体现足够光滑的曲面和边缘所带来的流畅效果。为了创造流畅的观览体验,运用了三维曲面来分隔空间,用非线性流线作为主要游览线路。用 AutoCAD 和 Rhino 及其插件 Grasshopper 软件来建立电脑模型,之后将 1:1 的图纸送到铁艺工厂,将 DWG 格式文件通过网络分别发送到模型公司及数控加工工厂进行材料切割。在完成焊接钢质材料之后,使用胶合和捆绑的方法将不同的材料结合在一起。

We chose the collar as our prototype because of its elegance and dignity, and we decided to use stainless steel in order to obtain an absolutely smooth surface and give clear edges to the curved surface to make a contrast between fluency and restraint. In order to create a fluent visiting experience, we use 3D curved surfaces to divide the space along a non-linear circulation route. We built the model using AutoCAD and Grasshopper. We then sent the 1:1 plan to a metal working factory and the DWG file to a model company to cut the plexiglass and the ABS file to a CNC factory to laser cut the stainless steel section. After welding, we glued and fixed the different materials together.

"球"椅 /
Sphere Chair

球椅和博物馆的设计思想来源于空间球形的切割和组合。通过组合不同的球,然后对球体进行布尔运算和空间演绎,得到最终的建筑和椅子形式。当然考虑到椅子和博物馆的不同,博物馆穹顶空间的设计为人提供了更多的游览乐趣。在获得最终形式之后,考虑了如何采用数控加工工具进行制作,将不同的形式分为不同的组,然后采用数控机床切割成弧形条状构件,再进行拼装及打磨,使椅子最终成型。

The design of the Sphere Chair was inspired by the cutting and combination of spherical spaces. By combining different spherical spaces, we carried out a Boolean operation on the volume, which became the prototype for both the building and the furniture. The difference between the chair and the museum is that the museum has a dome over its space and is more fun to visit. After deciding on the final form, we took into account how to fabricate the object using a CNC milling machine. We exported the files and translated them into code for to cutting, and then milled the wooden form in sections. Because the depth of drill, we need to sub-divide the surface. After we had finished cutting, we assembled and polished the form so that it became a chair.

"茧"椅 /
Cocoon Chair

以蚕茧为意象,层层抽丝。在实际建模过程中,我们运用 Grasshopper 运算生成椅子的编织面层,再把相同的运算逻辑运用于建筑的表皮。最终椅子依照程序逻辑通过手工编制完成,而建筑模型表皮则由 3D 打印技术实现;将手工、程序以及机械三个方式结合于设计之中,展现其不同的魅力。

Our project used the cocoon as a source of inspiration and attempted to express the feeling of weaving layer upon layer. In the process of modeling, Grasshopper was utilized for simulating the weaving of the surface patterns on the chair. The same logic was applied to the skin of the museum as well. During the process of fabrication, the chair was made by hand while the skin of the museum model was produced by 3D printing. Physical work, machine and computational simulation are shown in the different aspects of our project and give a sense of charm to the advanced digital techniques and craftmanship.

中国东南大学建筑学院 /Southeast

际村 /
Ji Cun

际村项目的核心是找到一种传统的居住空间模式的生成系统，同时解决规划和建筑设计两方面的问题。这一生成系统的核心则是从传统的居住模式中提取关于村落及其样式的自组织法则。生成系统提取建筑设计中可程序化的参数，寻找一种最合适的方式将计算机生成结构与基地现状结合起来。通过定义某些参数及其独立的属性和方法，设计程序逐渐发展成为一种基于多代理系统的、复杂的、可以实现自我优化的系统。

The reconstruction of the Chinese traditional village Ji Cun is the focus of this project which realizes a traditional style residential space generating system, addressing both urban planning and architectural design issues. The main concerns of the generating system are the self-organizing principles of the village and the patterns extracted from traditional residences. The generating system extracts programmable factors within the architectural design and explores the best way to combine the computer generated results with the current situation of the site. Through the definition of certain agents and their independent properties and methods, the design process turns into the optimization of a complicated self-adaptive system based on Multi-Agents Systems.

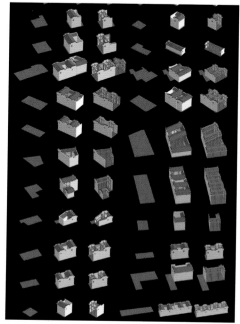

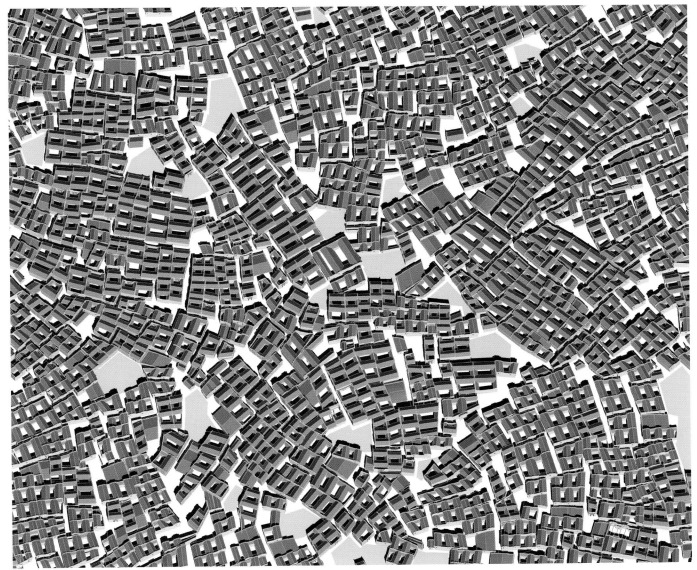

服务中心 /
Server Center

南京青奥服务中心项目是一个基于系统的数据链方法在建筑建造上的大规模尝试。设计加工过程均采用了由 JAVA 语言控制的 CNC 切割技术。为了获得每一块穿孔铝板安装的精确角度和板上孔洞的位置，我们在项目系统中建立了一种映射机制，在图片的像素信息与建筑表皮的空间信息之间建立关联。该项目的实施证明了数据链方法不仅可以应用于全局控制，也可以应用于加工制造过程。

The project of "Server Center of YOG, Nanjing" is a large scale proposal for architectural construction, based on the systematic "Digital Chain" method. The design and the fabrication process using CNC milling are controlled by JAVA (Processing) programming. A mapping mechanism which connects the pixel information and the spatial components of the architectural surface is established in this programming system, in order to get the information of the precise angle of folding and hole position for every different aluminum sheets. This project illustrates how the "Digital Chain" method makes contributions not only to the global optimization of the design, but also to Processing and fabrication at the same time.

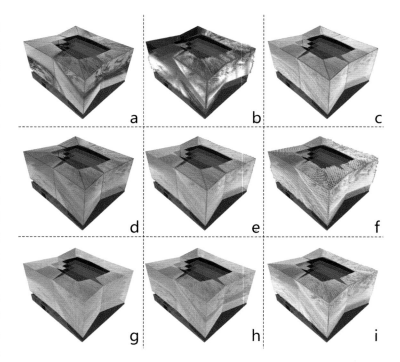

中国天津大学建筑学院 /Tianjin

互动光栅 /
Optical Grating

互动光栅希望探索一种通过变幻的视觉回馈与人产生互动的可能性，该项目利用了光的衍射原理和人类视觉暂留及补足的特性。这个系统需要使用经过处理的图底照片，搭配可以移动的有机玻璃格栅，通过格栅上下左右移动，在墙面上会产生连续变幻的动画效果。用 Arduino 控制板以及 Processing 程序设定，Sensor 及 Leap 感应人的指令，通过移动的光栅产生变幻的图像效果并与人产生互动。

The Interactive Architecture Optical Grating project explores the possibility of changing visual effects in response to human behavior. The project uses the principle of light diffraction and the characteristics of visual persistence and complementarity. The Optical Grating moves in all directions, collocating with disposed photos, producing continuous changing effects. A motion sensor device responds to human movements, producing an ever-changing interactive effect through the use of Arduino and Processing language.

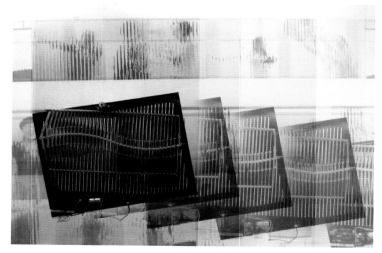

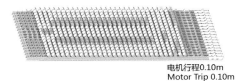

电机行程0.10m
Motor Trip 0.10m

电机行程0.20m
Motor Trip 0.20m

电机行程0.30m
Motor Trip 0.30m

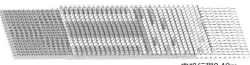

电机行程0.40m
Motor Trip 0.40m

水冰伏特加 /
Water, Ice, Absolut Vodka

针对伏特加文化的创意活动，我们创作了这套"水，冰，绝对伏特加"的互动装置。装置立意于伏特加的酒吧文化，强调酒本身的纯粹性及酒瓶的标志性。装置可感知人的行为，并通过内部机械设备将伏特加酒瓶的形状以多组透明亚克力块翻转的方式呈现。设计中我们使用电脑编程计算和模拟其内部机械结构，并在制作中使用了CNC，激光切割加工零件。而使用Arduino搭建的控制系统则能很好地实现装置与人的互动。

We designed this "water, ice, Absolut Vodka" interactive device in response to Absolut Vodka's creative activities. The device illustrates Absolut Vodka's bar culture, emphasizing the purity of the spirit and the symbolic shape of the bottle. The device can sense human behavior, and by using mechanical equipment, the shape of Vodka bottle is represented by the rolling acrylic blokes. In the design we used computer programming to calculate and simulate the internal mechanical structure; and we used CNC, laser cutting to fabricate the parts. The control system is designed using Arduino in order to achieve an effective interactive engagement with the users.

中国同济大学建筑与城市规划学院 /Tongji

反转檐椽 /
Reverse Rafter

本项目抽取了传统木构建筑中"檐椽"这一元素,希望通过新的优化结构性能模拟和机器人加工手段,对这一传统元素的结构性能进行优化与重新演绎。《营造法式》中对檐椽的出挑比例有明确的规定,我们通过基于参数化设计平台的结构计算软件 Millipede 和遗传算法优化运算器 Galapagos,对这一出挑比例进行了验算与优化。结合由三根杆件形成基本的三角形自支撑单元,设计了一套新的结构体系。机器人的精确加工能力能够很好地解决这些问题,使得本装置的结构意图得到完美呈现。

This project chose the "rafter", as an element from the traditional timber structural system, and attempted to reproduce and develop this structural element through structural simulation, optimization software and robotic fabrication. The overhang of a rafter had been clearly laid out in the Yin Zao Fa Shi traditional construction manual and we attempted to optimize it by using Millipede and Galapagos. Based on the results obtained from the optimization process we designed a new structural system incorporating a kind of basic self-supporting triangular component consisted of three timber members. Robotic fabrication offers a very accurate way of fabricating this, so that the structural logic of this installation can be demonstrated perfectly.

空间 // 空间 /
SPACE//SPACE

本设计试图处理题目提出的对立性与联系性。对于一个建筑来说，这种二元性存在于室内与室外，实体与虚空，具体功能与空间漫游。对于一个观演中心来说，这种二元性存在于封闭的观演空间与开放的城市空间，目标人群与周边居民。本设计选择了极小曲面这种建筑原型来处理这种二元性，二元之间实现相互咬合而又相互独立。

This design started with an attempt to deal with the opposition and the relations implicit in the brief. With buildings in general there is always an opposition that exists between indoors and outdoors, solid and void, and between specific constraints and the freedom to wander through space; with a theatrical building, this opposition also exists between the closed theatrical space and the open urban space, and between the theatre-goers and the nearby residents. For this design the logic of the minimal surface was chosen as the prototype to overcome this opposition and to allow both sides to interlock and fold into one another.

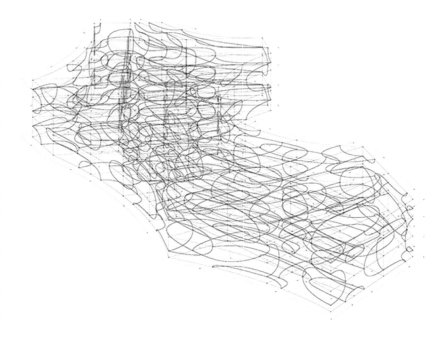

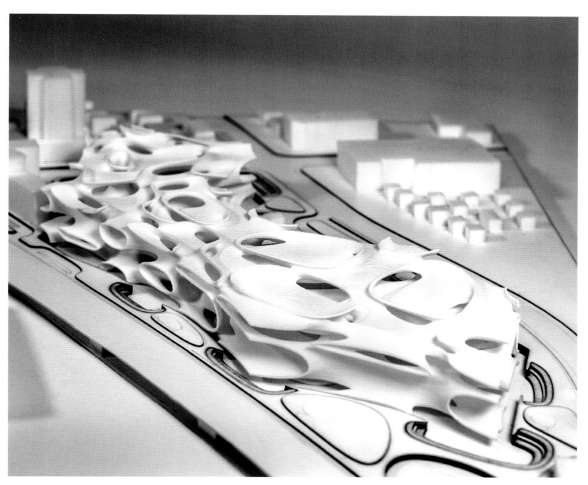
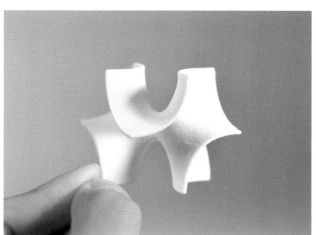

多面折叠 /
Folding Polyhedron

本设计从研究空间腔体开始，每一个腔体都是承载功能活动的单元。将菱形十二面体组合、堆叠，并提取相交棱线，得到如分子结构般的空间结构体系。在均质的体系内形成四种直纹曲面来控制空间的收放，对应博物馆各个空间对功能、流线的不同需求。最终形成的建筑系统既有多面几何体的规整秩序，又有直纹曲面的收放韵律。

This design started from a study of spatial volumes, where each volume is a unit of street activities. By stacking up rhombic dodecahedrons and extracting their edges, we can build a spatial structural system, which looks like a molecular structure. Four kinds of ruled surfaces can be formed in this system. They can satisfy different demands for program and circulation. As a result, the museum not only illustrates the logic of polyhedron geometries, but also shows the rhythm of ruled surfaces.

中国清华大学建筑学院 /Tsinghua

游牧空港 /
Nomadic Airport

本设计以"快速机场"为题目，通过功能复合和"智能汽车系统"的引入，探讨未来机场城市综合体的可能性。为了保证空间的高效和自由，引入了德勒兹"游牧空间"的概念，通过对遗传算法、人流模拟、DLA算法、Subdiv算法等的综合运用，实现了游牧空间的建筑转译。

This design, entitled "Nomadic Airport", explores the possibility of a future airport city complex by employing the concepts of a functional mix and a "smart car system". In order to make the space efficient and free, the design was based on the concept of "Nomadic Space" promoted by Gilles Deleuze. It then attempted to realize the architectural translation of "Nomadic Space" by the integrated use of genetic algorithms, human-flow simulations, diffusion-limited aggregation algorithms and subdivision algorithms.

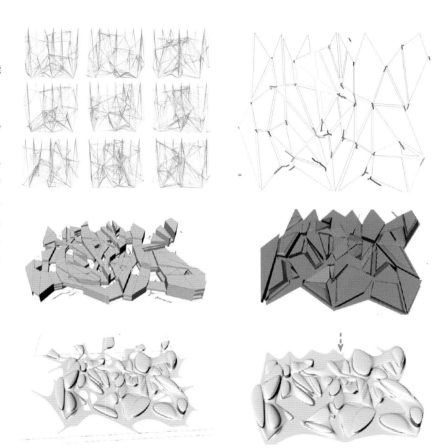

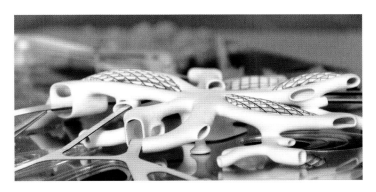

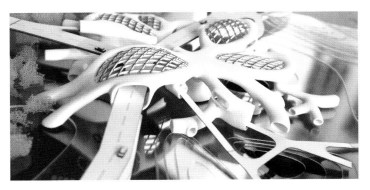

桥的城市 /
Bridge City

2011年3月，日本仙台遭受了严重的海啸与地震侵袭，导致沿岸损毁严重，当地政府计划将这个片区规划为纪念性景观公园。设计选取了海岸北部的一个候鸟栖息地作为设计区域，以恢复原先候鸟栖息地的风貌，并适当植入人的活动区域为出发点，提出了分形城市的概念。为保持纵横交错的水网形态，建立抵御海啸的屏障，区域的路网、建筑的生成方式选择了DLA和Branch两种枝杈形生成算法。

Sendai was one of the most seriously devastated cities in Japan after the tsunami of 2011. The memory of the disaster is still fresh. This project focuses on turning the damaged land into a memorial public park. With the desire to recover the landscape of the habitat for the migratory birds, to make this area a memorial landscape park, the concept of the 'fractal city' was selected. Diffusion-limited aggregation algorithms and L-systems were deployed to shape the organic forms for the buildings and to make the roads correspond to the maze of rivers and provide a shelter against the tsunami.

机器人建造 /
Robotic Fabrication

这组课题的要点是将人的身体语言数据采集下来，随后转换成空间坐标点数据库，并通过一系列算法转换，最终转化为一种艺术化的视觉表现形式。学生们运用微软的Kinect动作识别器，将人在5秒内的身体动作通过关节点的空间坐标录制下来，并建立了一个相对完整的空间点阵数据库。接下来，通过对这个数据库的梳理和算法的转换，在计算机内生成空间模型。最后，将空间模型转换为机械臂的加工路径，并根据加工的要求设计相对应的机器手，最终完成物理模型的生成。

The challenge of this group was to collect and convert human body languages to coordinate a database of spatial points, and finally transform them into an artistic visual presentation through a series of algorithmic translations. Firstly, one complete space lattice database was established according to the transcribed bodily movements of one person in the form of the spatial coordinates of articulation points using a Microsoft Kinect motion sensing device. Secondly, a spatial model was generated through the collation of this database and algorithmic translations. Thirdly, with the conversion of this spatial model to embody robotic arm processing procedures, and the design of a corresponding robotic hand, a physical model was finally generated.

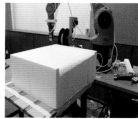
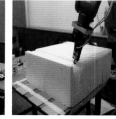

索引 / INDEX

America

CCA

Geoweaver
Jia Wu, Mary Sek,
Jeff Maeshiro
Tutored by Jason Kelly Johnson, Michael Shiloh

Stratum Networks
Max Sanchez, Taylor Fulton
Tutored by Jason Kelly Johnson, Michael Shiloh

Swarmscapers
Clayton Muhleman, Alan Cation, Adithi Satish
Tutored by Jason Kelly Johnson, Michael Shiloh

Columbia GSAPP

Fulfillment Center
Jonathan Yang
Tutored by Toru Hasegawa, Mark Collins

Pixels
Sucheta Nadig
Tutored by Toru Hasegawa, Mark Collins

Provisional Spaces
Yongwon Kwon
Tutored by Toru Hasegawa, Mark Collins

Harvard GSD

Light Forest
Wenting Guo, Juhun Lee
Tutored by Panagiotis Michalatos

(Re) Fabricating Nervi
Sofia Koutsenko, Javier Martin Fuentes
Johannes staudt, Juan Pablo Ugarte
Tutored by Leire Asensio Villoria

Synaesthetic Pavilion
Zheng Cui
Tutored by David Mah

MIT

Knitting Behavior
Carrie McKnelly
Tutored by Larry Sass,
Caitlin Mueller, Brandon Clifford

Making Gestures
Diego Pinochet
Tutored by Terry Knight

Learning From the Master
Guillermo Bernal
Tutored by Federico Casalegno

Pratt

Blurred Lines
Matthew Leta, Andrew Reitz
Tutored by Richard Sarrach, Ajmal Aqtash

Snow Angel
Leland Jobson, Andrew Reitz
Tutored by Holy Roller

Ign[o]rhizome
Keshav Ramaswami,
Hayden Minick
Tutored by Ezio Blasetti

Princeton

Light as Creative Medium
Dalma Földesi, Sonia Flamberg
Gina Morrow, Regina Teng
Tutored by Ryan Luke Johns

Reimaging Fabrication
Nicholas Pajerski, Gabriel Fries-Briggs,
Brendan Shea
Tutored by Axel Kilian

Tantalus' Hubris
Jeffrey Anderson, Tyler Kvochick
Tutored by Ryan Luke Johns

RPI

Methodological Hybrids
Graham Billings, Joseph Marchiafava
Tutored by Francis Bitonti

Pixilated Organicism
Emily Klein, Chris Muscari
Tutored by Adam Dayem

Objectified Alibis
Lisa Laue
Tutored by Brian De Luna

SCI-Arc

Self-Titled
Andrew Adzemovic
Tutored by Elena Manferdini

Surfaces to Flowers
Leonora Bustamante
Tutored by Hernan Diaz Alonso

Painterly Digital
Omer Pekin
Tutored by Elena Manferdini

UCLA

Swiss Army Knife Rooms
Timothy Mo Harmon
Tutored by Georgina Huljich

Porous Mass
Timothy Mo Harmon
Tutored by Georgina Huljich

The Aether Project
Refik Anadol, Raman Mustafa,
Julietta Gil, Farzad Mirshafie
Tutored by Guvenc Oze, Casey Rea

UMichigan

AL-10XX
Drew Delle Bovi, Grant Herron, Mark Knutson,
John Larmor, Shan Sutherland
Tutored by Glenn Wilcox

Domain
Ellen Duff, Stefan Klecheski, Jordan Lutren,
Ryan Scanlan
Tutored by Sandra Manninger

Agent Intelligence
Yinglin Wu, Safei Gu, Jiahui Wang
Tutored by Glenn Wilcox, Wes Mcgee

UPenn

Black Holes
Jaeho Jin
Tutored by Tom Wiscombe

The Unimaginable Library
Anyi Song, Xiaoqing Leah
Tutored by Cecil Balmond, Ezio Blazetti

INFINITEsimal
Jose Holguin, Harry Lam, Billy Wang
Tutored by Cecil Balmond, Ezio Blazetti

USC

Dailemota
Hanze Yu, Siyu Cui, Kaining Li
Tutored by Jose Sanchez

Alice in Wonderland
Justin Edwards, Corey Koczarski,
Kristina Lambros, Niccole Landowski,
Su Liu, Phillip Royster,
Paulina Shahery, Weiyi Tan, Wing (Vince) Tang
Tutored by Neil Leach, Myles Sciotto

LACMA++
Yuchen Cai
Tutored by David Gerber

Yale

Patternism
Constance Vale, Belinda Lee
Tutored by Brennan Buck

Carapace
Nguyen Lynn
Tutored by Greg Lynn, Brennan Buck

Disheveled Geometries
Junpei Okai, Dionysus Cho, Emily Bell,
Elvira Hoxha, Peter Le, Anne Ma, Daniel Nguyen,
Madelynn Ringo, John Wolfe
Tutored by Mark Foster Gage, Adam Wagoner

Europe

AA

Infusion
Soulaf Aburas, Giannis Nikas,
Mattia Santi, Maria Paula Velasquez
Tutored by Shajay Bhooshan

OwO
Agata Banaszek, Camilla Degli Esposti,
Ilya Pereyaslavtsev, Antonios Thodis

Tutored by Theodore Spyropoulos, Mostafa El-Sayed

Aerial Printed Infill
Duo Chen, Liu Xiao, Sasila.
Krishnasreni, Yiqiang Chen
Tutored by Robert Stuart-Smith, Tyson Hosmeri

Angewandte

Sematectonic Fields
Shajay Bhooshan
Tutored by Zaha Hadid, Mario Gasser
Christian Kronaus, Jens Mehlan, Robert Neumayr
Patrik Schumacher and Hannes Traupmann

Unfolding Monolith
Rangel Karaivanov
Tutored by Greg Lynn, Maja Ozvaldich
Bence Pap, Parsa Khalili

H+
Noemi Polo, Arpapan Chantanakajornfung
Tutored by Hani Rashid

Bartlett

SanDPrint
XiYangZi Cao, Shuo Liu, Zeyu Yang
Tutored by Daniel Widrig,
Stefan Bassing, Soomeen Hahm

Fibro City
Esteban Castro, Marcin Komar,
Aikaterini Papadimitriou, Yilin Yap

Tutored by Alisa Andrasek, Daghan Can

Filamentrics
Nan Jiang, Yiwei Wang,
Zeeshan Ahmed, Yichao Chen
Tutored by Manuel Jimenez Garcia, Gilles Retsin

CITA

Supple Architectural Surfaces
Olga Krukovskaya
Tutored by Martin Tamke

The Growth of Specificity
Lukasz Wlodarczyk
Tutored by Phil Ayres

Synthetic Felt
Asya Ilgün
Tutored by Jacob Riiber

TU Delft

Urban Interface
Jeroen van Lith, Mohammad Jooshesh,
Vasiliki Koliaki
Tutored by Henriette Bier, Sina Mostafavi,
Ana Maria Anton, Serban Bodea,
Matteo Baldassari, Vera Laszlo, Marco Galli

Vault
M. Augaityte, S. Daalmeijer, L. ter Hall
J. Johnstone, B. Kozlowski, A. Marcassoli
P. Pizzi, I. Plastira, A. Rossi, M. Schwartz
B. Tamasan, M. Ubarevicius, P. van den Hof

L. Vester, D. Zhou, M. Zucchi
Tutored by J. Feringa, M. Rippmann, S. Oesterle

Scalable Porosity
B. Raaphorst, G. Mostart, H. de Jonge,
J.M. van Lith, J. Paclt, K. Siderius, M. Galli,
M. Kornecki, M. Jooshesh, O. Anghelache, P. Low,
R. Flis, Rob W. Moors, R. Roodt, R. Hoogenraad,
S. Hoeijmakers, S. Kanters, T. IJperlaan
Tutored by H. Bier, S. Mostafavi,
A. Anton, S. Bodea

DIA

Poseidon Pods
Alexandru Trifan,
Ramy Maher
Tutored by Alexander Kalachev, Karim Soliman

Float Project
Kavinaya Sakthivelan,
Angela Bogdanova
Tutored by Alexander Kalachev

Firefly System
Mostafa Seleem, Nicholas Canlas
Tutored by Alexandeer Kalachev, Karim Soliman

EGS

Synthetic Ecologies
Nicole Koltick
Tutored by Ben Bratton, Neil Leach

Healing the Urban Tissue
Omar Henry
Tutored by John Frazer

0°N
Alex Webb
Tutored by Neil Leach

IAAC

KATA
Agnieszka Wanda Janusz, Kateryna Rogynska,
Alessio Salvatore Verdolino,
Tobias Grumstrup Lund Øhrstrøm,
Tutored by Enric Ruiz-Geli, Mireia Luzárraga
Maria Kupstova, Dori Sadan

Translated Geometries
Efilena Baseta, Ece Tankal, Ramin Shambayati
Tutored by Areti Markopoulou,
Alexandre Dubor, Moritz Begle

TRANS[X]PONENTIAL
Aditya Kadabi, Shweta Das, Boney Virendra
Keriwala
Tutored by Willy Müller,
Pablo Ros, Jordi Vivaldi

Paris Malaquais

A Degree of Freedom
Jim Rhone
Tutored by Philippe Morel, Pierre Cutellic

RO-BORIE
Marie Lhuillier
Tutored by Philippe Morel, Chirstian Girard

Additive Manufacturing
Romain Duballet, Clement Gosselin, Philippe Roux
Tutored by Philippe Morel, Jean-Aimé Shu

ICD Stuttgart

Augmented Grain
Dylan Marx Wood
Tutored by O. Krieg, D. Correa, A. Menges

Encoded Surfaces
Boyan Mihaylov
Tutored by Achim Menges, Jan Knippers

Carbon Fiber Pavilion
Moritz Doerstelmann, Vassilios Kirtzakis,
Stefana Parascho, Marshall Prado, Tobias Schwinn
Tutored by Achim Menges, Jan Knippers

ETH Zurich

Sequential Frames
David Jenny, Jean-Marc Stadelmann
Tutored by Fabio Gramazio, Matthias Kohler,
Michael Budig, Raffael Petrovic

Mesh Towers
Petrus Aejmejlaeus-Lindstrom,
Pun Hon Chiang, Ping Fuan Lee
Tutored by Fabio Gramazio, Matthias Kohler,
Michael Budig, Raffael Petrovic

Vertical Avenue
KaiQi Foong, Lijing Kan
Tutored by Fabio Gramazio, Matthias Kohler
Michael Budig, Raffael Petrovic
All projects: Gramazio Kohler Research,
SEC Future Cities Laboratory, 2013

Australia

RMIT

Ibrosity
Cam Newnham, Adam Pehar
Tutored by Roland Snooks

SmartNodes Pavilion
Brendan Knife, Joshua Salisbury-Carter,
Max McCardle, Prachi Lai
Haomin Zhang, Zheng Guo
Tutored by Kristof Crolla,
Nicholas Williams, Jane Burry

The Spatial Field
Vicki Karavasil
Tutored by Vivian Mitsogianni

China

CUHK

HK Cultural Centre

LAM, Yue Shing Chandler
TSUI, Sze Man Eunice
Tutored by Kristof Crolla

Gothic Reconstruction

Wan, Ka Kwan April
Chan, Wing Chun Frankie
Tutored by Kristof Crolla

Bamboo Grid Shells

Chan Wai Kan Wilkin, Chi To Leung Mark,
Chan On Ki Ok, Cheng Man Lai Karen,
Du Qiongwei Winnie, Chan Hansen,
Ho Chun Sing Jason, Wai Hon Lam Joshua,
To Hon Yin Ernest, Yam Ka Kit Kevin,
YIM Yu Ching Ben, Ho Yuen Ling Elaine
Tutored by Kristof Crolla, Ip Tsz Man Vincent,
Julien Klisz, Goman Ho, Alfred Fong,
Jia-Jin Wang

HKU

Media Tecture

Eunice Faith Asis
Tutored by David Erdman

Museum Boom

Felix Chu Shun Lok, Adrian Luk
Tutored by Tom Verebes

Systems of Multiplicity

Xiang Wang
Tutored by Christian Lange

Nanjing UA

Body Without Organs

Rui Sun, Zhiwei L, Chang Liu,
Renjie Dai, Zhencheng Xu
Xiaoyuan Sh, Xuhua Zhang
Dekui Kong, Xueyuan Wu
Xiaohui Gu, Shanshan Li, Qingqing Li
Tutored by Feng Xu, Lei Yu,
Jiong Xu, Ke Zhen Wang

Digital Forest

Lingchen Kong, Yujie Yang, Dekui Kong
Wang Lu, Jianjun Li, Pu Xu
Yue Yang, Peishiqiu Ping
Tutored by Jiong Xu, Heping Zhan

Multidimensional Rockery

Yu Wu, Shanliang Chen, Guohui Zeng
Yangxiao Wu, Xiaoying Wang, Mi Chen
Qiang Hu, Yujie Yang, Xingle Cai
Huizhong Ding, Huaqing Zhang
Tutored by Jiong Xu, Heping Zhan

SCUT

Collar Chair

Lu Lu, Xinyu Xiao
Tutored by Gang Song

Sphere Chair

Mu Qiao, Hongxuan Zeng
Tutored by Gang Song

Cocoon Chair

Yi Sun, Yingxin Wu
Tutored by Gang Song
Yinghe He, Minghui Xiao

Southeast

Ji Cun

Zifeng Guo, Yunzhu Ji
Tianyue Shi, Jiaqing Cao
Tutored by Biao Li, Hao Hua

Server Center

Zifeng Guo
Tutored by Biao Li

Tianjin

Optical Grating

Ba Jing, Zhang Zexi, Li Qiyuan,
Liu Lu, Shen Tao, Zhao Mengjie,
Li Tian, Liu Chengming,
Lei Huanlingzi, , Gao Qifeng,
Lv Lifeng, Qi Shan,
Wang Yingying, Zhang Aoyu,
Yang Hui, Ge Kangning
Tutored by Xu Zhen, Michael Fox

Water, Ice, Absolut Vodka

Su Chong, Li Yunxiang
Tutored by Xu Zhen

Tongji

Reverse Rafter
Zhang Liming, Li Xiuran, Chai Hua

Tutored by Philip Yuan,

Panagiotis Michalatos

SPACE//SPACE
Shi Ji, Liu Xun, Li Ximeng

Tutored by Philip Yuan

Folding Polyhedron
He Yiwen

Tutored by Philip Yuan, Lyla Wu

Tsinghua

Nomadic Airport
Lu Shuai, Zhao Yizhou

Tutored by Xu Weiguo, Huang Weixin,

Shen Yuan, Jordan Kanter

Bridge City
Li Xiaoan, Li Ruiqing

Tutored by Xu Weiguo, Liu Nianxiong,

Huang Weixin, Wanghui

Robotic Fabrication
Li Le, Ma Yao, Sun Chenwei,

Gao Yuan, Wang Zhi, Wu Xiaohan

Tutored by Xu Weiguo, Yu Lei

作者简介 /BIOGRAPHIES

徐卫国 /Xu Weiguo

徐卫国教授执教于清华大学建筑学院，现为建筑系系主任；曾是美国麻省理工学院访问学者，2011~2012年执教于美国南加州建筑学院（SCI-Arc）及南加州大学建筑学院（USC）；并曾在日本留学，获日本京都大学博士学位，工作于日本村野藤吾建筑事务所。他在任清华建筑教授的同时，创立XWG建筑工作室，从事建筑设计工作，并在多项设计竞赛中获奖。他是国内参数化非线性建筑设计的开拓者及领军人，主持多项国家自然科学基金项目的研究。发表论文80余篇，出版专著及编著11本。1999年参加第20届UIA国际建筑师大会中国青年建筑师作品展，之后又在世界各地讲演或举办展览；他还是北京国际青年建筑师及学生作品展（2004、2006、2008、2010）的策展人；2012年作为主要发起人，创立数字建筑设计专业委员会并被选为专委会主任；2013年组织DADA系列活动。

Xu Weiguo is Professor and Chair of Architecture in the School of Architecture at Tsinghua University. He was a Visiting Scholar at MIT in 2007 and taught in SCI-Arc and USC in 2011-2012. He studied architecture at Tsinghua University, and then started teaching at the same institution before moving to Japan to work for Murano Mori Architects. He was awarded his doctorate from Kyoto University in Japan. On returning to China, he established his own architectural practice (XWG) in Beijing. He is the recipient of many awards. He is a leading scholar and architect in the digital design field in China. And he has taken charged of the reseach projects funded by National Natural Science Fundation of China, including " Theory and Method of Digital Architecture Design (2015) ","A research on the Method of Parametric and Non-linear Architecture Design (2010) ","A research on the Generating and Fabricating of Non-standard Architectural Form (2005) " His 80 works have been published in many journals. He is the author of 11 books including The Way of Architectural Design Thinking (China Architecture and Building Press, 2001), Architecture/Non-Architecture (China Architecture and Building Press, 2006), Studio Works of Tsinghua Students (Tsinghua University Press,2006), and co-author of Fast Forward>>(MAP Books, 2004), Emerging Talents, Emerging Technologies (2 Vols., China Architecture and Building Press, 2006), (Im)material Processes: New Digital Techniques for Architecture (2 Vols., China Architecture and Building Press, 2008), Machinic Processes (2 Vols., China Architecture and Building Press, 2010), Design Intelligence Advanced Computational Research DADA (2013). Xu Weiguo was included in the Exhibition of Young Chinese Architects at UIA Congress 20th in 1999, and was selected as one of the architects to represent China in the A1 Pavilion at ABB 2004. He was one of the curators of Architecture Biennial Beijing 2004, 2006, 2008 and 2010. He has had lectures and exhibitions all over the world. As one of main initiators, he established the Digital Architecture Design Association and was elected Director of DADA in 2012. He organized the DADA series events in 2013.

尼尔·林奇 /Neil Leach

尼尔·林奇是一位建筑师兼理论家，他目前是欧洲研究生院的教授，哈佛大学以及同济大学的访问教授，以及南加州大学建筑系的副教授，同时他还是美国国家航空航天局的高级创新专员。他曾经执教于南加州建筑学院、英国建筑联盟建筑学院、康奈尔大学、哥伦比亚大学建筑学院、西班牙加泰罗尼亚高级建筑研究学院、德国德绍建筑学院、巴斯大学、布莱顿大学和诺丁汉大学，其著作包括《空间政治》（劳特利奇出版社即将出版）、《伪装》（麻省理工学院2006年出版）、《忘掉海德格尔》（派迪亚出版社2006年出版）、《中国》（香港麦普奥菲斯出版社2004年出版）、《千年文化》（伊利普西思出版社1999年出版）和《建筑麻醉学》（麻省理工学院2006年出版）；曾与他人合著《玛思潘滋》（建筑基金出版社2000年出版）；曾编辑《空间建筑——设计研究的新前沿》（怀利出版社2014年出版）、《数字城市》（怀利出版社2009年出版）、《为数字世界而设计》（怀利出版社2002年出版）、《空间的象形文字》（劳特利奇出版社2002年出版）、《建筑与革命》（劳特利奇出版社1999年出版）和《建筑的反思》（劳特利奇出版社1997年出版）；与他人共同编辑《集群智能：多代理系统建筑》（同济大学出版社即将出版）；《人机未来》（同济大学出版社2015年出版）、《探访中国数字建筑设计工作营》(同济大学出版社 2013 年出版)、《建筑数字化编程》（同济大学出版社 2012 年出版)、《建筑数字化建造》（同济大学出版社 2012 年出版)、《数字建构：青年建筑师/学生设计作品》（两册，中国建筑工业出版社 2010 年出版）、《涌现：青年建筑师/学生设计作品》（两册，中国建筑工业出版社 2006 年出版)、《快进·热点·智囊组》（香港麦普奥菲斯出版社 2004 年出版）、《数字建构》（怀利出版社 2004 年出版）；还是《阿尔伯蒂建筑十书》（麻省理工学院 1988 年出版）的译者之一。他目前正在参与由美国国家航空航天局赞助的研究项目，此项目着眼于开发可以用在月球及火星上的三维打印结构的智能机器人。

Neil Leach is an architect and theorist. He is a Professor at the European Graduate School, Visiting Professor at Harvard University GSD and Tongji University, Adjunct Professor at the University of Southern California, and a NASA Innovative Advanced Concepts fellow. He has also taught at SCI-Arc, Architectural Association, Cornell University, Columbia GSAPP, Dessau Institute of Architecture, Royal Danish School of Fine Arts, IaaC, ESARQ, University of Bath, University of Brighton and University of Nottingham. He is the author of The Politics of Space (Routledge, forthcoming), Camouflage (MIT Press, 2006), Forget Heidegger (Paideia, 2006), China (Map Office, 2004), Millennium Culture (Ellipsis, 1999) and The Anaesthetics of Architecture (MIT Press, 1999); co-author of Mars Pants: Covert Histories, Temporal Distortions, Animated Lives (Architecture Foundation, 2000); editor of Space Architecture: The New Frontier for Design Research (Wiley, 2014), Digital Cities (Wiley, 2009), Designing for a Digital World (Wiley, 2002), The Hieroglyphics of Space: Reading and Experiencing the Modern Metropolis (Routledge, 2002), Architecture and Revolution: Contemporary Perspectives on Central and Eastern Europe (Routledge, 1999), and Rethinking Architecture: A Reader in Cultural Theory (Routledge, 1997); co-editor of Swarm Intelligence: Architectures of Multi-Agent Systems, (Tongji University Press, forthcoming); Robotic Futures (Tongji University Press, 2015), Design Intelligence: Advanced Computational Research (China Architecture and Building Press, 2013), Digital Workshop in China (Tongji University Press, 2013) Scripting the Future (Tongji University Press, 2012), Fabricating the Future (Tongji University Press, 2012); Machinic Processes, 2 Vols. (China Architecture and Building Press, 2010), (Im)material Processes: New Digital Techniques for Architecture, 2 Vols., (China Architecture and Building Press, 2008), Emerging Talents, Emerging Technologies, 2 Vols., (China Architecture and Building Press, 2006), Fast Forward>>, Hot Spots, Brain Cells (Map Office, 2004) and Digital Tectonics (Wiley, 2004); and co-translator of Leon Battista Alberti, On the Art of Building in Ten Books (MIT Press, 1988). He is currently working on a research project sponsored by NASA to develop a robotic fabrication technology to print structures on the Moon and Mars.

图书在版编目（CIP）数据

数字工厂　高级计算性生形与建造研究：学生建筑设计作品／徐卫国，（英）林奇编. —北京：中国建筑工业出版社，2015.12
ISBN 978-7-112-18912-0

Ⅰ.①数… Ⅱ.①徐…②林… Ⅲ.①建筑设计—作品集—中国—现代 Ⅳ.① TU206

中国版本图书馆 CIP 数据核字（2015）第 300176 号

责任编辑：张　建
责任校对：李欣慰　关　健

数字工厂
高级计算性生形与建造研究
学生建筑设计作品
徐卫国　尼尔·林奇（英）编
*
中国建筑工业出版社 出版、发行（北京西郊百万庄）
各地新华书店、建筑书店经销
北京盛通印刷股份有限公司印刷
*
开本：889×1194 毫米　1/20　印张：11$\frac{1}{5}$　字数：396 千字
2015 年 12 月第一版　2015 年 12 月第一次印刷
定价：98.00 元
ISBN 978-7-112-18912-0
（28188）

版权所有　翻印必究
如有印装质量问题，可寄本社退换
（邮政编码 100037）